"Don't rock the boat.
This is an order."

Bury Him

A Memoir of the Viet Nam War

Captain Doug Chamberlain

Copyright 2019 by Doug Chamberlain.
Published 2019.
Printed in the United States of America.
All rights reserved.

No portion of this book may be reproduced, stored in a retrieval system, or transmitted in any form or by any means – electronic, mechanical, photocopy, recording, scanning, or other – except for brief quotations in critical reviews or articles, without the prior written permission of the author.

ISBN 978-1-950647-03-3

Publisher's Cataloging-in-Publication data
Names: Chamberlain, Doug, 1942. author.
Title: Bury him / Doug Chamberlain.
Description: Parker [Colorado] : Bookcrafters, 2019. Hardcover.
Identifiers: ISBN: 978-1-950647-05-7
Subjects: LCSH: United States. Marine Corps—History—Vietnam War, 1961-1975. | United States. Marine Corps—Officers—Biography.
BISAC: BIOGRAPHY & AUTOBIOGRAPHY / Military
Classification: LCC E840.5 | DDC 920 CHAMBERLAIN–dc22

The front cover photograph of Marine Corporal Joseph T. Wnukowski was furnished courtesy of Corporal Wnukowski and Marine Corporal Joseph W. Freda, who took the picture.

Love the West Publications
Publishing Services by BookCrafters, Parker, Colorado.
www.bookcrafters.net

For most of his life, Captain Doug Chamberlain has lived and worked on his family ranch in LaGrange, Wyoming. Doug has had experience in public education as both a classroom teacher and administrator, in radio broadcasting, and the transportation industry. He has been a community and state leader serving in the Wyoming State Legislature for 18 years culminating in his election and service as Speaker of the Wyoming House of Representatives.

Perhaps his greatest public service, however, was his time as a Marine Company Commander in Viet Nam. It was a time for him to serve his country, but at the same time, the source of one of his greatest regrets. Like many Viet Nam vets, Doug went over alone and came home alone. And, like many other vets, he came home with baggage, violent memories and unfinished business. He was haunted by what he did and failed to do in the war.

In this frank, engaging memoir, Captain Chamberlain casts a long look back to his service in the Marines. He chronicles the missions, personal courage and sacrifice of the Marines he was privileged to command; painfully recalls the unspeakable order he and his Marines were forced to obey; and the cover-up which followed.

Nearly four decades later, Captain Chamberlain makes right what was wrong; brings closure to the family of a fallen and abandoned warrior; and attempts to put to rest the guilt which plagued his military career and life.

Unlike most books on the Viet Nam War, this book is written at a tactical level by a Marine Company Commander who was there. Unlike most books on the Viet Nam War, this book is written at a tactical level by a Marine Company Commander who was there. Serious historians, Viet Nam War scholars and military commanders at all levels and branches will laugh, cry and identify with the lessons of war found in Chamberlain's memoirs as he writes about the indelible connection between Marines in combat.

Readers will marvel at Captain Chamberlain's tenacity in uncovering and ultimately correcting a lie and highlighting the phony heroes, weak politicians, liars, and fabricators who tarnished the reputation and sacrifice of the men and women in uniform who fought and died in Viet Nam.

A must read for young military leaders who think the only enemies in war are those combatants on the other side of the FLOT (Forward Line of Troops).

Major General Edward Wright, United States Army (Retired)

Captain Doug Chamberlain's book reaffirms the stark reality, daily agonies and the constant vigilance of commanding a company of Marines and Corpsmen in combat during the height of the Vietnam War in 1968. It is both story and history describing threatening events with split-second decisions changing lives, forever! This book is a compelling read, honest, up-close and personal told by a Company Commander who took his responsibilities seriously! "Skipper" Chamberlain, not only "tells the rest of the story" but "tells the real story" of his time in the field with his Marines and Sailors and of the traumatic, life-changing ordeal of being ordered by higher-ups to bury a U.S. Marine KIA....

Grady T. Birdsong, Corporal USMC - 1/27, 2/9 & 3rd MarDiv Communications Company, RVN 1968-1969.
Author of *To the Sound of the Guns*

In war, the only way to preserve one's morality is to apply a strict moral code to the conduct of war itself. Captain Doug Chamberlain was one of the longest serving Marine Company Commanders in the Vietnam War. His honor, and that of the Marine Corps, was tested severely when he was ordered to bury the body of a Marine instead of bringing home the remains. Bury Him *is the story of a commander's commitment to his Marines and his determination to redeem his honor and that of the Corps.*

Don R. Catherall, PhD.
Author of *The Handbook of Stress, Trauma and the Family*
and *Shaped by the Shadow of War*

I have known Doug Chamberlain to be a man of constant character and integrity since I first met him as a student athlete at John Brown University in the 1960s. His book recounts his agonized response to a direct order to "bury" the remains of a fallen Marine in Vietnam, followed by a prompt retraction at Doug's urging and a heroic recovery and return of that body to the family. However, the confusing and critical response from a few officers in the Marine Corp Command set Doug upon a difficult 50-year journey of unanswered questions and great personal stress. His final victorious discovery of the "truth, the whole truth, and nothing but the truth" is a gripping and tough personal narrative that will inspire every reader in difficult circumstances to speak truth, regardless of the consequences.

John E. Brown, III
Past president of JBU, and former AR State Senator

Captain Chamberlain's searing memoir of combat in the southern I Corps zone, during the height of the Vietnam War, not only tells gripping stories of heroism and tragedy... it also unveils, for the first time, the truth of a sorry episode in Marine Corps history that has remained buried for fifty years. Chamberlain's earnest prose reveals a burdened conscience, but it also demonstrates his unflinching courage in fulfilling his duty as a Marine and a patriot, ultimately proving himself a warrior with his honor intact. I was privileged to help him uncover documentary evidence of the events at the heart of his story, bringing closure after half a century for a company of Marines who were asked to do the unthinkable, and for a family who never understood why they had to bury a loved one—not once, but twice. Bury Him *can stand proudly alongside such enduring classics of Marine Corps literature as Robert Leckie's* Helmet for My Pillow *and Philip Caputo's* A Rumor of War.

Paul T. Semones, P.E.
Semones Forensic Engineering

Dedication

This book is dedicated to all United States Marines and their families; Doug Berger; the memory of PFC Michael J. Kelly; the Marines who were serving in Echo Company, Second Battalion, Seventh Marines on 25 March 1968; and everyone else who has ever served in the Armed Forces of the United States of America. Semper Fi.

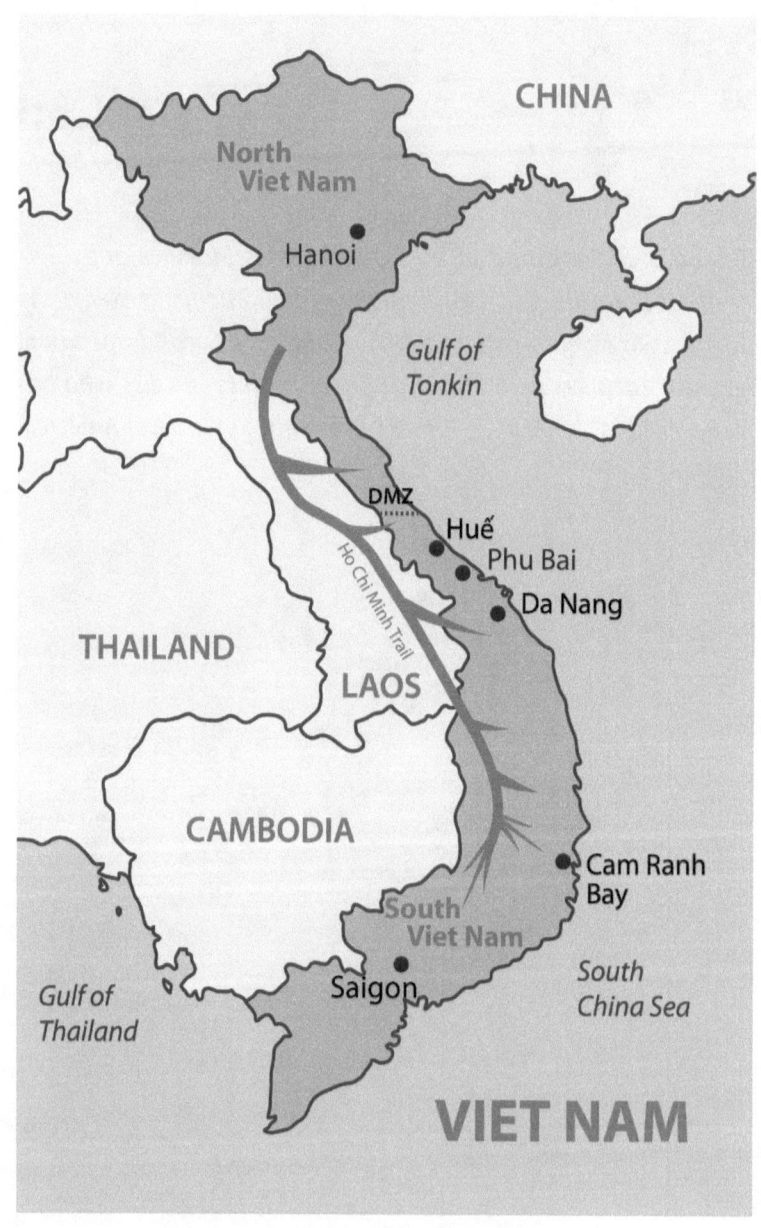

Map of Vietnam and the surrounding countries with Ho Chi Minh Trail, Map courtesy of Nick Zelinger, NZGraphics.com

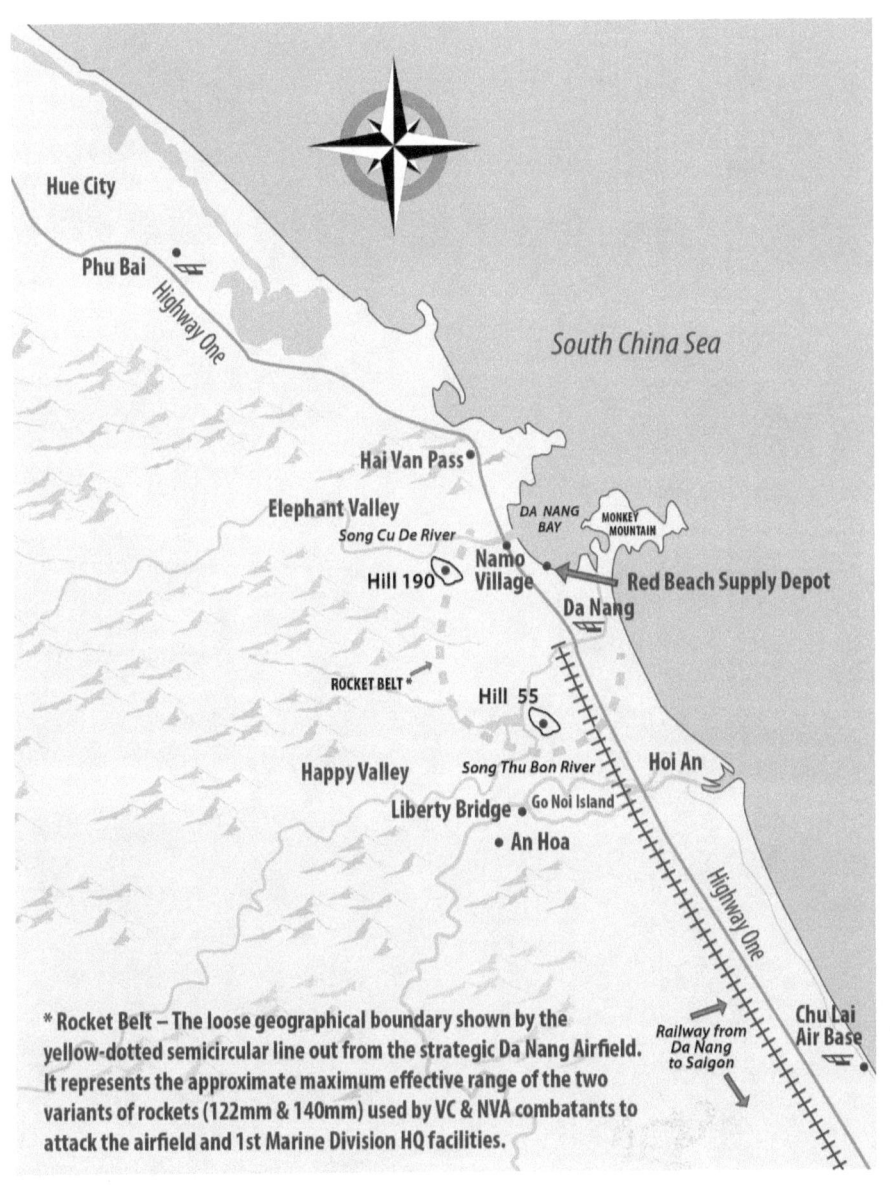

"Echo Company, 2nd Bn, 7th Marine Regiment Area of Operations 1967-68
Map courtesy of Nick Zelinger, NZGraphics.com

...THE SEATS TO DENVER BRONCO FOOTBALL GAMES have been sold out for so many years that no one can remember when they weren't. I have been a season ticket holder since the late 1990s. This particular Sunday afternoon was no exception. Over 76,000 Bronco fans rose to their feet at the Sports Authority Field at Mile High. They stood to honor the American and organizational flags being carried by the Color Guards entering the stadium through the north end zone. The United States Air Force, the United States Army, the United States Coast Guard, The Denver Fire Fighters, The United States Marine Corps, The United States Navy, and the Denver Police Department were all represented. The small contingents moved briskly to the yard markers on the east side of the field to which they were assigned, pivoted to their right, and then more deliberately and ceremoniously followed their yard line across the field. They stopped in a parallel line near the boundary line facing the sea of humanity standing in the west side of the stadium. The Marine Corps Color Guard was standing on the 50-yard line. In the usual sequence, the Stadium Announcer introduced the contingents beginning in order from the north end of the stadium. The roar of the crowd followed each introduction. I heard him say, "The United States Marine Corps"... and the sounds from the stadium began to fade away from me. I felt an overwhelming, smothering wave of shame that seemed to crush my chest and I thought my head was going to explode. Suddenly, I was standing alone in a bomb crater in the jungles of South Viet Nam. I was looking down at the partially decomposed remains of a Marine that appeared to have been abandoned and left behind. I felt myself shaking, and someone was approaching me with their hand out...and when they touched my left arm, I froze. I quickly looked at the person and soon realized it was a woman standing beside me who had purchased the

Bronco game ticket for the seat next to mine. She was rubbing my arm and appeared to be looking at me with concern and compassion, but she said nothing. I had no idea who she was before the game, and I never had the opportunity to see her again after that day...

Table of Contents

Preface..1
Introduction..5
The Candidate...17
Basic School and Basketball...26
Destination: Okinawa..34
In-country..37
2nd Battalion, 7th Marine Regiment, 1st Marine Division................39
Learning the Actual Military Tactics and Operations of a War........41
"Go Get 'em, Tiger"..48
The "Grapevine"...60
Aid and Comfort to the Enemy..64
Artillery and Fixed-Wing Support...67
Promotion..71
Elections..74
Hill 190..76
A Dream Came True...81
Man's Best Friend...87
Cool, Clear Water...93
Moving on with the Mission...95
"Thou Shalt Not Steal. Thou Shalt Not Kill"...................................104
The Military Industrial Complex...114
Chain of Command...120
Propaganda..124
Winning the Hearts and Minds of the People.................................127
"Huey"...138
"wbp"...140

Operation Pitt	144
Christmas, 1967	150
Tet	152
Leech Valley	165
Ice Cream?	168
Noncombat Casualties	170
Unsung Heroes	172
Lives Changed Forever...Operation Worth	174
The Rest of the Story	187
Operation Ballard Valley	195
Operation Allen Brook	202
"Mother's Day Massacre"	212
R&R	224
Special Landing Force	227
Special Landing Force Deployment	231
Rotation Date	233
Reporting as Ordered	235
Civilian Life and POWs/MIAs	241
The Search for Truth	251
The Truth, the Whole Truth, and Nothing but the Truth	253
Semper Fi	271
The Betrayal of American Youth	275
Post-Traumatic Stress Disorder (PTSD)	280
In Retrospect	285
Epilogue	288
Personal Biographical Information	293
Attachments	299
Acknowledgements	332

Preface

THE UNITED STATES MARINE CORPS is, and has been, the finest military force the world has ever known. One of the reasons is because of the esprit de corps that is created in those who serve in the capacity of being a United States Marine. The sacrifices endured in the training, and the expectations of those who have earned the title "Marine," sometimes transcend the understanding of those who have never "been there and done that." Most of the finest people I have ever known, and with whom I have been associated with, are United States Marines. The circumstances under which we became known to each other could play a part in my evaluation of these great Americans, but I have compared them to the standards of the hundreds of people I have known in public service and the business ventures in my life, and my dedication to these Marines has only been reinforced. With that said, any organization, team, or military branch is only as strong as its weakest member. There are always some in any common entity that fail to meet the standards of the larger group, and the results can be devastating, but not necessarily fatal overall.

Many of the names of the people in this book are individuals I was associated with once I arrived in South Viet Nam. Some of the names are fictitious to protect the identity and privacy of the individuals, and their families. However, the exceptions are Sergeant John "Jack" Johann; Lieutenant Caesar "Sid" Stair, III; Corporal Donald "Cathi" Catherall; Lance Corporal Gabriel "Gabe" Quinonez; Staff Sergeant Lutu; Lieutenant Jimmie R. Pippen; Corporal Joseph W. Freda; Corporal Joseph T. Wnukowski; Lieutenant Colonel Charles Mueller; and Lieutenant Colonel John R. Love, who was deceased in 2010. In addition, Captain Lyle Johnson's name is listed as Killed In Action in Viet Nam in public documents. The other names in the military records after my Forensic Investigation was initiated are as they appear in those documents.

I could not have performed my duties as the Commander of Echo Company, Second Battalion, 7th Marine Regiment, First Division of the United States Marine Corps with any kind of success if it had not been for the very courageous, professional, dedicated, patriotic, outstanding Marines in Echo Company, Second Battalion, Seventh Marines. My debt of gratitude to them can never be repaid. The word "hero" has been attributed to many in our society today. However, the word has special meaning to me when it is used in reference to anyone who has risked their own life and personal safety for the assistance to, and the defense of, the well-being, safety, and lives of others.

In my particular case, the Marines who served under my command of Echo Company, 2nd Battalion, 7th Marines will always be my heroes, and I will never forget the sacrifices we endured. As you read my accounts of the things I saw through my eyes, please keep in mind that every good and successful thing that happened to Echo Company while it was under my command was as a result of the never-ending support and efforts by all of us to make our unit succeed. I have intentionally left out many of the details of their individual work, in most cases, because I do not believe I can accurately describe the heroic efforts of my men or relate what they were thinking throughout our time together. I can only describe what I saw, and this book is my attempt to articulate the reasoning behind the decisions I made. After enormous thought and years of personal deliberation, this is my attempt to put into words what I recall during this period of my life from my perspective.

One perspective of my writing needs to be clarified as it relates to using the term "gook" in reference to the enemy forces in South Viet Nam. The term is used in a majoritive sense in the historical context of this writing. It also appears to have been used in the Korean War by American troops as nomenclature to reference the enemy. I fully understand that it can be considered to be an ethnic slur in today's interpretation, but that is not what is intended in this book. The term is only used as a historical reference to the terminology that was used by American forces during the Viet Nam War. We also referred to those in that same general group as "Charlie." In a similar manner, I include the use of the terms "baby killers," "murderers," "rapists," and Marine "dogs" that were used by our Vietnamese enemies and the anti-war activists in the United States and around the world to describe our American Military personnel during that time. I would appreciate the tolerance and latitude by you, the reader, to allow me to preserve the historical accuracy of my book in this manner.

The final impetus for me to commit myself to the effort necessary to write this book came as a result of a conversation I had with Dr. Brent Kaufman, DVN, in 2015. Dr. Kaufman is the co-owner of the Goshen Veterinary Clinic in Torrington, Wyoming. He was a student of mine when I taught at La Grange High School in La Grange, Wyoming, after my release from active duty and following my first attempt to study jurisprudence at the University of Wyoming. I was also his basketball coach during those years at LaGrange. On the day of the conversation, Dr. Kaufman was administering medical treatment to one of my cattle, and he and I were jokingly reminiscing about those days at LGHS when he suddenly became serious. He said he had always wondered why I always seemed "mad and angry" during those years. That was the first time that I could ever remember anyone asking me about my behavior during that period of my life who thought it was abnormal, without being judgmental. He was not comparing me and my conduct to what and who I was prior to my military service, because he did not know me then. He was simply making an observation. His comment haunted me for several months, and it eventually led to the writing of this book.

Introduction

My "roots" are near the edge of the eastern Rocky Mountains in southeastern Wyoming and western Nebraska. My fraternal grandfather came to southeast Wyoming from northeast Indiana to live with relatives in the late 1800's and to seek natural relief from the effects of tuberculosis, according to family historical accounts. It was while he was here that he homesteaded property east of what is now the town of La Grange. Family records indicate that my father and his four siblings were born on the homestead property. My father was born in 1909. Our family records indicate that the actual occupying of the Chamberlain homestead property began prior to 1900. The actual ownership document was called a "Patent" that could be obtained from the federal government after residing on a 160-acre parcel of land continuously for at least 10 years. Additional land was added to the original 160-acre parcel and I still live on that original homestead.

My mother was born in 1912 on one of the Warner homestead farms in extreme western Nebraska approximately 10 miles east of where my father was born on the homestead in Wyoming. My maternal grandparents were descendants of a large, extended family that emigrated from Sweden. The first historical record of the family involved Andrew Warner in Newtowne, Massachusetts, which later became the city of Cambridge. At what appears to be the first record of a town meeting held on December 24, 1632, he was given the responsibility of building "20 rods" of fence "to enclose the common," a task similar to that assigned to many other settlers in Newtowne.

As we say in rural Wyoming and Nebraska where I was born and raised, "my roots go way back and are very deep." Both of my parents' families were poor by any standard, but both families survived by knowing how to be good stewards of the land, excellent practitioners of animal husbandry, and most of all, being great friends and neighbors.

I am the youngest of five children, and I was born in 1942. I had two Great-Uncles who served in the United States Army in World War I, and both were physically and/or emotionally disabled to some degree as a result of their service. One was my Great-Uncle Roy Hamilton, who was from my mother's side of the family. He related to me that he had spent 43 continuous days in the trenches that were tactically used during his service in his infantry unit. "Uncle Roy" was plagued by continual "twitching" in his shoulders and torso, and he continuously tugged lightly at his clothing. My Great-Aunt Alice, his wife, said he had been like that since returning from "the War." However, he lived until he was 90, and she lived to the "ripe old age" of 102.

My other relative who served in WWI was my Great-Uncle Guy Hanna, who also served in the United States Army Infantry as a Scout. He was from my father's side of the family. He returned with severe damage to his lungs from exposure to mustard gas, and he had overwhelming emotional problems as a result of his service. Family memories of him were that loud noises would make him very nervous, and for several years, when an airplane would fly over, he would dive onto the ground, cover his head with his arms and hands, and shout warnings for everyone to "get down." My siblings who remember him said that he was very withdrawn, and that he would often sit in a darkened corner of a room by himself. They reminisced that his behavior was so abnormal that some of their teenage friends were afraid of him. He was born 50 years to the day before I was. A bunkhouse where he lived near my grandparent's home after his return from "the War" is still standing here on the homestead, although it is in a very serious state of deterioration. I have resolved to myself to try to preserve it, but I have never had the resources to be able to do it to this point in my life.

I had an Uncle Oris Chamberlain, who served in the United States Army in World War II in a Construction Battalion, and whose son, my first cousin, Oris Chamberlain, Jr., served in the United States Army Infantry in the Korean War. Junior, as we called him, was seriously wounded shortly after his arrival in Korea, and he bore the physical disabilities associated with those injuries for the remainder of his life. He recalled that when he finally regained consciousness after being hit by rounds fired from a machinegun, someone had removed his boots and taken them for their use because they thought he was dead.

My brother, Earl W. Chamberlain, served in the United States Army after completion of the Reserve Officers Training Corps (ROTC) Program and his graduation from the University of Wyoming in 1954. He was stationed in Germany for a period of time during part of the "Cold War Era."

Growing up in rural Wyoming was a "storybook" life. My parents created our sustenance by farming and ranching. We never went to bed hungry, but there were meals every day that were the creation of a mother who was very inventive and an excellent cook. There were few people, if any, that could be considered wealthy near the town of La Grange, the population of which was approximately 150 people. My education was obtained in a school system where there were only 30 students in the high school. My graduating class consisted of 10 young Americans, the largest class in the history of the school up until that time. We were taught that we could contribute to the land of the free and the home of the brave. It was instilled in us that we should try to make a better life "for ourselves and those whose happiness depended upon us," which was part of our Future Farmers of America (FFA) youth organization pledge. My childhood was one where respect for others and dependency on others for survival were an important key to the military leadership that would be required of me later in life, even though I could not imagine it at the time.

From this small school and town in Wyoming over a four-year period came four state basketball championship teams in a row, four college athletes, one NCAA basketball coach, one banker and financier, two military Officers who served in Viet Nam, one Enlisted military personnel who also served in Viet Nam, some ranchers, some teachers, and some business owners, just to name a few of the accomplishments of our small group of students.

We did not have kindergarten in our small school, so some of us entered school when we were five years old. Our inspiration for learning began with Martha Ann Adams, our first and second grade teacher, who was one of the kindest persons I have ever known. My inspiration to become a pilot began with the boundless efforts of my third and fourth grade teacher, Shirley Mills. She was an unmarried teacher in a small town in Wyoming and was what adult males in our little town considered a "Knockout." At our young age, we thought all of our teachers were beautiful, but what made Miss Mills particularly outstanding was her ability to relate to us and make us believe in ourselves, even if we wanted to be a pilot. She taught us all of the very simple fundamentals of flight. I eventually became a private pilot and had the tremendous pleasure of having her fly with me from Wyoming to southern California for an educational convention some 20 years later. We learned vocal music from Janette Kessler, a wonderful, but firm, Vocal Music Teacher who inspired us to love music and taught us how concentrated effort was essential to succeed in life. She was a fabulous mother, a great ranching partner with her husband, and an outstanding role model to all of us. Cleo Wheeland was a traveling maestro who came to our school twice a week and taught us instrumental music so some of us could play in the school band. Marching and playing in county fair parades were summer activities that he always demanded of us. Two coaches, Bill Perich and Ron Schliske, taught us the fundamentals of basketball. At the same time, they taught us the life lessons of self-discipline, personal sacrifice, and competing to win fairly while respecting our opponents. They required us to condition ourselves to develop endurance and instilled in us the courage to never quit, regardless of the difficulties we faced. Because of the teaching, training and inspiration of two high school Vocational Agriculture Teachers and Future Farmers of America Advisors, Marvin Riley and Jack Lester, I was elected to be the State President of the

Wyoming Future Farmers of America during my senior year in high school. I was awarded an American Farmer Degree because of the efforts of these kinds of great people that played such an important part in shaping my life during my childhood and adolescent years. Little did I realize at the time that the lessons they taught us would play a key role in the discipline required in my future military experience.

One of my classmates, Robert "Robby" Greathouse, became an Officer in the United States Navy, after graduating from Colorado State University (CSU). He attended CSU where he had been awarded a full track scholarship. He "walked on" and played football for the CSU Rams, also. Robby flew F-8 fighter jets and bombers for two tours in Viet Nam from August of 1968 to February of 1969, and from August of 1969 until February 1970 aboard the USS Hancock. After serving our Country for 25 years, he retired from the Naval Service as a Navy Captain. Captain Greathouse then pursued a career as a commercial airline pilot and retired from Alaska Airlines after 12 years of service to that organization. I was also a military officer, and served as an infantry officer in the United States Marine Corps.

Another classmate of ours, Robert "Greg" Gregory, was a distant relative of mine because one of his uncles on his father's side of his family was married to one of my aunts on my father's side of our family. Greg enlisted in the United States Army, and his service in Viet Nam was concentrated on the technological monitoring of the performance and capabilities of the Russian built SAMs, or Surface to Air Missiles, deployed in North Viet Nam. In addition, he and his unit also had the ability to monitor the developing Soviet intercontinental ballistic missile program that was later deployed in the Soviet Union during the Cold War Era. The Soviets used the rockets that they developed during that time in the early stages of the Soviet Space Program. The practical application of his duties in Viet Nam also included the operation of drones for collection of intelligence information while flying over North Viet Nam. Marine Infantry Units were used to help recover the cameras from those drones that sometimes crashed into the jungles upon their return to South Viet Nam. After completing his enlistment, Greg became employed as a civilian to develop more sophisticated drones that were to be used by the United States in the future. His work, and that of his colleagues, laid the foundation for drone technology being used by the United States military today.

I was fortunate enough to receive an athletic scholarship to play basketball and run track at the University of Wyoming. Even though I was enrolled in the ROTC during my freshman year at UW, military service had never really occurred to me. After leaving UW, I transferred to John Brown University, or "JBU," in Siloam Springs, Arkansas, where I continued my education to become a teacher. The cost of my education was paid for with a scholarship I received for playing basketball. While there, I was honored to be elected Student Body President. JBU is a private, nondenominational University that was founded by a Methodist Evangelist, John Brown, Sr. Upon my graduation in May of 1964, I returned to Wyoming to teach and coach basketball, track, and cross country at Veteran High School in Veteran, Wyoming.

Teaching and coaching provided me with the ability to make monthly payments on a new 1964 Corvette. However, after one carefree year of "living the dream," my life unimaginably took a turn down a road of no return. I received a notice from the United States Selective Service Office to report for induction into the United States Army in the spring of 1965. I had been drafted. The draft lottery had not been initiated in those early years of the "Conflict." Draft selections were determined by a draft board that consisted of persons in each county in every state who were appointed by the President of the United States. In reality, young people of draft age eligibility who did not have any political connections or financial "where-with-all" seemed to be the first that were selected, and I was one of them. Flexibility was allowed as far as to which branch of the service a draftee was to enter if the draftee wanted to enlist, and I thought I would like to learn how to fly and be a "jet fighter pilot." A slight vision deficiency in one eye kept me from pursuing that. Since I was a college graduate, I reviewed the programs in all branches of the military that were available for potential Officer Candidates. The Officer Candidate Training Program for the United States Marine Corps was the shortest, so I notified the Wyoming Selective Service Office in Goshen County that I was enlisting in the United States Marine Corps. After all, it would only be a few weeks of training beginning October 17, 1965.

The sudden change in my plans for my life also presented me with a decision as far as possible marriage plans. I had dated one young lady for most of my high school and college years. After I received my draft notice, we decided that it would be best if we were married before I left to fulfill my military obligation. Even though being married at that time was a way to be exempt from the draft, I did not avail myself of that excuse to avoid serving my Country. Hindsight being "20-20," marriage at that time was probably a mistake. We could not, and did not, anticipate the life-changing experiences that awaited me. In fairness, after my military experience, I am sure I was not the person my wife thought she had married.

I was not anxious to interrupt my "budding career" as a high school basketball coach. Sports had always been a significant part of my life. However, my family heritage included military service, especially when our Country called. I knew very little about Viet Nam, primarily because I did not think about possible military duty in Southeast Asia at that time of my life. When the reality of being drafted finally struck me, I began to listen to the media coverage of what appeared to be the United States becoming involved in a land war in that part of the world. The United States government had been publicly repeating a mantra about the spread of Communism in South Viet Nam, Laos, and Cambodia, and how it was endangering Thailand, Burma, and Indonesia. Some reports maintained that eventually even Australia could possibly be directly affected. As a result of these predictions, the leadership in our Country at that time steadfastly insisted that U. S. military intervention in South Viet Nam was necessary to slow, and eventually deter, the advancement of this atheistic, dictatorial "scourge." Some historical prognosticators through the years have suggested that the assassination of President

John F. Kennedy could possibly have been as a result of his determination that our Country should not become engaged in a land war in Asia, and in particular South Viet Nam. Whatever the case, President Kennedy had been "taken out," and Vice President Lyndon Johnson became President and Commander in Chief of the United States Military Forces after President Kennedy's tragic death on November 22, 1963.

The run-up to the drafting and enlistment of several million young Americans for service in the Viet Nam War was punctuated by an alleged attack on United States Navy ships in the part of the South China Sea called the Gulf of Tonkin, which is just off the coast of North Viet Nam. What now appears to have been the propaganda of the Johnson Administration was that on August 2 and 4, 1964, several relatively small North Vietnamese gunboats attacked our Navy fleet operating in international waters in that Gulf. It was admitted by the Johnson Administration at the time that there was very little damage, if any, to our naval vessels, and some credible military sources at the time claimed that the attacks did not even take place. The President consulted with his Secretary of Defense Robert McNamara, Secretary of State Dean Rusk, National Security Advisors, and Joint Chiefs of Staff. Their collective decisions eventually led to the commitment of American military aircraft and as many as 525,000 ground troops during 1968 in the effort to provide the South Vietnamese "self-determination." In their wisdom, the alleged "attacks" were to have been so egregious that our Country's chosen response led to the longest war in our nation's history up until that time.

In the ensuing years, hundreds of American pilots were killed conducting bombing raids over North Viet Nam, and hundreds more were captured and held as prisoners of war, or POWs, by the North Vietnamese. One of the pilots who was shot down and became a POW was John McCain, who later became a United States Senator from Arizona. POW McCain was the son of a United States Navy Admiral at the time he was shot down and captured. Without a doubt, he became the most recognizable of the many POW pilots. What later became the obvious "insanity" of the conduct of the Viet Nam War can best be illustrated by the following excerpt from a book that includes details of an alleged interview with the aforementioned Secretary of State Dean Rusk.

The general conversation of the interview was the issue of why so many American planes were shot down during the Viet Nam War. The source of the information was disclosed to me in early March of 2018. It came from a fellow Marine Officer, Captain Ron Maines, who is also a John Brown University graduate. Ron is the epitome of a great person, and he is a friend of mine who was a CH-46 helicopter pilot. He served in the Viet Nam theatre in 1969 and 1970.

Ron referred me to *The Secret War and Other Conflicts* that was published in 2014 by General John L. "Pete" Piotrowski, ISBN 978-1-4931-6187-4 (h). The author rose from the rank of a Basic Airman to become a Brigadier General in the United States Air Force. The following is a quote from pages 246/247 of his book:

> *Nearly twenty years later, former Secretary of State Dean Rusk being interviewed by Peter Arnett on a CBC documentary called,* **"The Ten Thousand Day War."**
>
> *Mr. Arnett asked, "It has been rumored that the United States provided the North Vietnamese government the names of the targets that would be bombed the following day. Is there any truth to that allegation?"*
>
> *To everyone's astonishment and absolute disgust, the former Secretary responded, "Yes. We didn't want to harm the North Vietnamese people, so we passed the targets to the Swiss embassy in Washington with instructions to pass them to the NVN [North Viet Nam] government through their embassy in Hanoi." As we watched in horror, Secretary Rusk went on to say, "All we wanted to do is demonstrate to the North Vietnamese leadership that we could strike targets at will, but we didn't want to kill innocent people. By giving the North Vietnamese*

advanced warning of the targets to be attacked, we thought they would tell the workers to stay home."

No wonder all the targets were so heavily defended day after day! The NVN obviously moved as many guns as they could overnight to better defend each target they knew was going to be attacked. Clearly, many brave American Air Force and Navy fliers died or spent years in NVN prison camps as a direct result of being intentionally betrayed by Secretary Rusk and Secretary McNamara, and perhaps, President Johnson himself.

I cannot think of a more duplicitous and treacherous act of American government officials. Dean Rusk served as Secretary of State from January 21, 1961 through January 20, 1969 under President John F. Kennedy and Lyndon B. Johnson.

The following two sentences were then attributed to Mr. Arnett by the author:

Mr. Peter Arnett opined that "This would be a treasonous act by anyone else." A very sad revelation.

I hope the information in General Piotrowski's book that is alleged to be true will help you, the reader, achieve a more clear understanding of the intensely emotional contents of this book. To make the cost of the war more poignant, records indicate that the United States lost **3,744 fixed-wing aircraft** and **5,607 helicopters** in the conflict. The number of aircraft and American pilots that were lost as a direct result of the Johnson Administration's unbelievable practice of advanced target notification is unknown. The release of the documentary entitled "**The Vietnam War**" in 2017 by Ken Burns and Geoffrey C. Ward has been positively acclaimed by many critics, but I had not yet felt emotionally able to view it at the time of this writing. Marine Veterans that I know who have seen the series, and that I have spoken to personally, seemed united in their impression that the production is one that maximized the mistakes of the United States participants at all levels, while minimizing the aggressive, murderous acts of the Viet Cong and North Vietnamese, especially those committed against the South Vietnamese populace.

Information attributed to the National Archives and Records Administration indicate that at least **58,267 American lives were sacrificed** for our Country in this incomprehensible War, and their names are engraved on the Vietnam Memorial Wall as accounted for. Wounded Americans, both physically and psychologically, can be estimated to be in the hundreds of thousands, considering that an estimated 2.5 million Americans served in the Viet Nam War. Statistical information presents even more stark demographic information. According to a retired Marine Sergeant Major, of the total of those Americans killed:

39,996 of those killed **were 22 years old or younger**
8,283 were **19 years old**
31,695 were **18 years old**
12 were **17 years old**
5 were **16 years old**
1 was **15 years old**
8 were **women**
997 were **killed on the day they arrived in Viet Nam**
1,448 were **killed on the day they were leaving Viet Nam**
31 sets of brothers were **killed**
3 sets of fathers and sons were **killed**
Most Americans killed on one day was **245**
    on January 31, 1968
Most Americans killed in one month was **2,415**
    in **May 1968**
POW/MIAs still unaccounted for – 1,910

Two other countries contributed significantly to the war effort. **60,000 Australians served** in the Viet Nam War of which **3,000 were wounded** in action, and **521 lost their lives**. **South Korea** also made a valiant attempt to prove their allegiance to the United States effort. In doing so, they **committed 300,000 troops** to the War effort of which **10,962 were wounded**, and **5,099 were killed in action** trying to defend the South Vietnamese.

There is a "saying" among some Veterans that is simplistic, but direct: "If old men had to fight the wars, there wouldn't be any."

The Candidate

OFFICER CANDIDATE SCHOOL, or OCS, for the Marine Corps was, and I believe still is, located at Quantico, Virginia. I arrived on a train the evening of October 16, 1965, along with several hundred other aspiring Candidates. It was very similar to what has been portrayed in movies with a Drill Instructor, commonly called a "D.I.," exercising their lungs as they shouted profanity to impress these "college girls" with descriptions of how tough the next several weeks of their lives were going to be... and they were right. Most of those outstanding D.I.'s had limited formal educations, but they knew more about living and dying than most Candidates would ever know. Looking back, I do not remember anything that the D.I.'s told us at any voice level that was not true.

Marine Corps Base, Quantico, Virginia, was also shared with the Federal Bureau of Investigation for training purposes, but we did not have any interaction with them that we were aware of. Every Candidate was a college graduate except for a few meritorious enlisted Noncommissioned Officers who, because of their excellent service record, were given the opportunity to become a Commissioned Officer. We called these Candidates who had come up through the ranks "Mustangs." They were generally older than us "college punks," and far ahead of us as far as combat training was concerned. The World War II era barracks where we were quartered were on the immediate east side of the main railroad line that stretched between New York City, New York, and Miami, Florida. The part of the base we were located on extended east to the Chesapeake Bay. The Railroad divided the Quantico base making it mandatory to cross it for almost everything involved with our training. The tracks were so close to the barracks that at night the headlight from an approaching train looked like it was going to come through the squad bay, not to mention the noise and vibration as the train passed by. Even the "Mess Hall" was located on the west side of the tracks, which gave us ample opportunity to stand at attention while trains passed and our "chow" was getting cold. Because we marched to and from the Mess Hall, no talking was allowed, but the noise of the trains would sometimes allow for some "criminal mischief" in the eyes of the D.I.'s. The food was edible, usually, and the rule was "take all you want, but eat everything you take." Anyone caught throwing food away was required to eat it out of the garbage can, so that did not happen more than once. One morning I forgot to take milk for my dry cereal, and when I went back to get some, the milk was all gone. I learned a valuable lesson while I was eating the dry cereal during the ten-minute period we were allowed to eat...in silence.

Every Candidate was scored in three categories: 25% of the score was based on Leadership; 25% of the score was based on Academic Achievement; and 50% was based on Physical Achievement. Examples of the physical scoring were the Obstacle Course, the Confidence Course, pugil stick fighting, bayonet fighting, calisthenics, and trail runs. The Obstacle Course was a timed event, but the Confidence Course was not. The Confidence Course involved climbing over objects as high as 40-50 feet without the use of safety nets. A Candidate once asked a D.I. what would happen if one of us fell off one of the obstacles, and the answer was simply, "You won't." The trail runs were designed to make the Candidates stay as close as possible to the Officer leading the run, so that when he stopped and turned around to write names on his notepad, you did not get a written "chit" for lagging behind. The chit was a negative performance evaluation, and an accumulation of too many chits could cause the Candidate to be "washed out" of the program. If you washed out of the program, you did not return to civilian life. The failed Candidate was sent to Parris Island, South Carolina, to the Marine Corps Boot Camp where they would be trained for a two-year enlistment, starting out as a "buck private."

The service obligation for successful Officer Candidates was for three years from the date of the successful completion of OCS and attaining the rank of Second Lieutenant. All officers were considered Reserve Officers initially, and later the Marine Corps selected the Reserve Officers they wanted to augment to become Regular Officers. I was never aware of what percentage of Reserve Officers that were selected for augmentation. I declined my offer to become a regular officer that I received shortly after my graduation from OCS.

"Hazing" of Candidates by D.I.'s was not supposed to be physical or dangerous. However, the line was blurred. The purpose of most hazing was to make a very identifiable point and to observe the reaction of the Candidates. It started at 0400 hours (4:00 A.M.) on the first day in the barracks when a large aluminum garbage can was thrown down the squad bay by the D.I. while the other D.I. pounded on the lid of the can with a piece of pipe. Amidst it all was an enormous amount of profanity and references to our family heritage, while none of the Candidates knew what they were supposed to do or where to go.

The expectations of the Candidates was to do everything exactly the way we were told to do it, with no exceptions. Much of the training was designed to make sure we paid attention to detail, and we were scored accordingly. Immense pressure was maintained on Candidates as yet another way of scoring their progress, with limited time to do everything. One of the timed tasks was the time allowed to relieve ourselves. Not only did we have limited time, but the facilities were very limited in size and number, and urination and defecation was not allowed anywhere other than in the "head," or restroom. The initial exposure to the pressure caused a Candidate who had graduated Magna Cum Laude from Georgia Tech to go AWOL, or Absent Without Leave, after only being there three days. He was sent to Parris Island when he was located and arrested.

One detail requirement was that we were not allowed to wear our government issued plastic wristwatch outside for physical training in the mornings. On one particular morning I forgot to do that. After removing our "blouse" and folding it a certain way and placing it on the ground in a certain spot, we began our calisthenics. At one point I heard a soft voice behind me say, "Did you forget something, Candidate?" Nothing more was said until we had our nightly "rack drill" before the D.I. "tucked his babies in for the night." All of the other Candidates were required to stand at attention while I pushed my wristwatch the full length of the 50-foot squad bay and back... with my nose, while I was berated with profanity by the D.I.

Rack drills were sometimes more physical than others. One in particular would begin with all of the Candidates standing at attention on each side of our respective steel bunk beds, clad only in "skivvies" and a T-shirt. The D.I. would shout, "In the rack!" All Candidates were to be lying in their respective bunks at attention before the D.I. counted to three. Invariably, the Candidates on the bottom bunks were hitting their heads on the bunk above, and the ones on top were hitting their shins on the steel frame that was under their mattress while trying to jump in. Because the count was always fast, it was impossible to accomplish the command, and at the end of the count, another shout followed, "Out of the rack!" The exercise would be repeated until the D.I. decided we had endured enough "training."

On one particular night, when a muscular 200-pound candidate tried to jump into the top bunk, the small bolts that secured the top bunk to the frame on the bottom bunk broke, and the Candidate fell to the floor with the steel frame hitting his lower back as it fell on top of him. The Navy Corpsmen were called and he was taken to a medical facility. But...it did not happen during "rack drill" because "we did not have rack drills"...and every Candidate would confirm that if questioned about it. We learned some time later that the injured Candidate had been medically discharged from the Marine Corps because of the injuries to his back.

After "Lights out!" and a short night of sleep, each Candidate was required to be standing by their bunks with a bed sheet over each shoulder and their pillowcase over their head when the lights came on at reveille, which was at 0430 hours the next morning. Alarms and waking devices were strictly prohibited, and we had 15 minutes to make up our bunk beds, shower, shave, dress, and be standing outside in formation at attention. One purpose of that particular requirement was for us to develop inventive ways of "beating the system," and we did, since strict adherence to the rules was impossible. I found that exercise to be very beneficial when I was in the reality of a command in South Viet Nam.

Another requirement was to always lock our footlockers and wall lockers every time we left the squad bay. While we were standing at attention at the railroad crossing one evening, a Candidate whispered that he had forgotten to lock his footlocker, and word traveled through the platoon. When we were dismissed from formation at the barracks after we had eaten, we entered the squad bay to see what might have happened to his footlocker. To his relief, it appeared he had escaped any retribution until he opened the locker. His pressurized can of shaving cream had been used to write "Surprise Asshole" on the inside of the lid, and all his equipment in the locker was covered with shaving cream.

Dealing with injuries was an expectation, not an excuse to fail to complete duties. On one occasion, we were "bivouacked," or camped out, with the other three platoons in our OCS Company in a training exercise. Everyone was required to have their shelter half attached to the shelter half of the Candidate who was their bunkmate back in the squad bay. It made a small tent with the use of two small stick poles, some short metal stakes, and a few short pieces of small cord. We had to be dressed down to skivvies only and be sleeping in our sleeping bag.

At approximately 0400 hours, wild outbursts of obscenities awoke us as the D.I.'s ran through the area kicking all of the stakes out of the ground that were holding the shelter halves. Profanity laced shouts ordered us to get in formation, creating as much confusion as possible. As you can imagine, trying to find your "gear" and get dressed in complete darkness was a "cluster" beyond belief, and getting all your gear in your pack and then be standing in formation seemed next to impossible. Numerous Candidates had two left or right boots on feet that were often missing their socks. After the screams and shouts subsided, what might be imagined as a formation of the platoons stood at attention in the darkness. The Company Commander told the Platoon Commanders that whoever had their entire Platoon back in the barracks first would receive a free case of beer.

Since the barracks were three or four miles away, no one could guess what was going to happen. We were running in some sort of formation in the darkness and passing another platoon on a narrow dirt road when I heard the Candidate immediately behind me in the formation fall to the ground, shouting in pain. We were not allowed to stop, and we had to get to the barracks so our D.I.s could inspect our gear to see if we had all our equipment. Any Candidate who did not have all their equipment was issued a chit. The injured Candidate did not arrive until around 1000 hours, when he finally hobbled into the barracks. An examination later indicated that he had broken his ankle, but he had been ordered to continue to walk several miles back to the barracks with the injury. Rumors were that he, too, was medically discharged. We were the first Platoon to the barracks, but since all of us were not there at the same time, our Platoon Commander did not win the case of beer. While this kind of hazing might appear to be juvenile, it was one of many ways we were taught that there was a consequence for mistakes...and mistakes in South Viet Nam could cost Marines their lives.

After all the training, hazing, injuries, and "chits," only 32 of the 50 Officer Candidates in my Platoon graduated and received the single brown bar indicating we were Second Lieutenants in the United States Marine Corps. One Candidate who was highly regarded as a person by all of us could never meet the physical qualifications. He should never have been recruited because of the immense amount of weight he had to lose, and even though he lost 40 pounds during the several weeks of training, he failed the final physical test. My work ethic instilled in me from my agricultural background and my competition in athletics up to that point in my life was a considerable advantage. It led to more opportunities for me in the months ahead.

After being in OCS for three weeks, we were given "liberty" to leave the base on Saturday and Sunday. However, before we could be released on Saturday mornings, every member of each respective platoon had to pass a rifle inspection while we stood at attention in formation. One Candidate from Texas in my Platoon was notorious for not having a weapon that would meet the inspection requirements of the D.I. On one particular Saturday morning, a D.I. from another Platoon was inspecting our rifles. When he took the rifle from the Candidate from the "Lone Star State," he stared down the bore, or the barrel. As part of the inspection procedure, the Candidate had to have the bolt of the weapon locked open so the person inspecting the rifle could look down the full length of the barrel to see if it was clean.

As the D.I. calmly gazed down the barrel, he said, "Did it snow last night, Candidate?" My Texas friend shouted, "No, Sergeant!" After a pause, the D.I. said, "That's a shame. Now we can't track down the bear that shit in your barrel." Instead of returning the rifle to the Candidate, he very calmly jammed the rifle barrel into the grass-covered ground where we stood, forcing dirt into approximately four to six inches of the barrel. A few minutes later, we were notified that our Platoon had failed the inspection and we were dismissed from the formation. As we took one step backwards, pivoted, and left the area, the only thing left standing was the lone rifle with the barrel stuck in the ground. We were not allowed to assist with the cleaning of the rifle, but you can imagine how difficult it was to remove all that dirt from the barrel. Once the Candidate had spent over an hour digging the dirt out of the barrel and cleaning the rest of his rifle, "The Eyes of Texas Shown Brightly Upon Us," and we headed for the nation's capital for about 36 hours away from the rigorous training.

The "scuttlebutt" was that Washington, D.C. was the place to go. A mystical organization supposedly existed in the D.C. area called the Junior Officer's Protective Association, or JOPA. JOPA held "mixers" every weekend where military personnel, even as lowly as Marine Officer Candidates, could go to dance, drink, party, and meet women. The only restrictions were that we were not to break the law, not get arrested, not drive anyone else's vehicle, and to be back to the OCS barracks no later than 1800 hours on Sunday. In the group I hung out with, I was the only one who was married and the only one who did not drink alcohol. Therefore, I was always the designated driver. The excuse we planned to use if I was ever caught driving another Candidate's vehicle was that it was better for me to drive a vehicle that belonged to someone else than for someone to drive drunk. As far as the rules, many Candidates had the attitude that breaking rules was not all that important because, "What are they going to do to us, send us to South Viet Nam?"

One memorable night we ended up in an apartment belonging to some young women that one in our group had met that night at the Speakeasy Bar. Our Candidate from Texas who historically failed inspections, became very intoxicated...and sick. One of the women who lived in the apartment and I decided we needed to assist the drunken Texan, and we laid him on a bed to hopefully give him some relief from his nausea. No sooner had we laid him down than he vomited all over himself and the bed. Needless to say, I had the dubious duty of assisting this gracious woman in cleaning up the "alcoholic discharge" that reeked throughout the apartment. Little did I know that her acquaintance and the cleaning up of a drunk, fellow Marine would be significant later in my life.

One rule we were taught was "Marines never leave anyone behind." The training was explicit that there was never a reason or excuse to even consider doing it. To leave a fellow Marine behind was considered to be the most dishonorable thing any Marine could do, and that principle was the very foundation of our organization...no excuses, never considered, and it must never happen. Marine Corps pride, tradition, and honor demanded strict allegiance to that history of commitment and dedication to fellow Marines. The training incident when the candidate broke his ankle seemed to be in violation of this honored Marine Corps tradition.

The Officer Candidate Graduation was an impressive day. Our families were able to attend, and my parents and my wife came from Wyoming to see the festivities. At the conclusion of the ceremonies, the Marine

Corps tradition at that time was that each Officer Candidate would "flip" their D.I. a silver dollar when the D.I. gave them their Second Lieutenant bars. My parents were able to locate enough silver dollars in Wyoming for everyone in my Platoon, and I think we were the only Platoon that was able to honor our D.I. that way. That was one of the few times I ever saw our D.I. smile.

My "overall percentage grade" in the 38th OCS class was 92.11%. That made my class standing 32nd out of the approximately 498 Officer Candidates who entered the program. I do not know what part those scores and/or that rating played in any of my future duty assignments, but it is a part of my Marine Corps Officer Qualification Records.

Basic School and Basketball

AFTER THE COMPLETION OF OFFICER CANDIDATE TRAINING, we were given two weeks leave prior to the beginning of Marine Corps Officer Basic School, and I returned to Wyoming for the 1965 Christmas and New Year Holidays. In January of 1966, my fellow Officer Candidates and I, who had just become "Brown Bar" Second Lieutenants, entered Basic School. In the movie "Patton," which was supposed to be the true story of United States Army General George Patton's military life, he is portrayed delivering a speech to the West Point Cadets. In the speech, he supposedly said, "The object of war is not to die for your country. The object of war is to make sure some other poor dumb bastard dies for his." We were now supposed to learn how to try to make that happen. In our training, the object of war was defined in a less glamorous way. We were simply told that "The object of war was to kill people and break things."

United States Marine Corps Officer Basic School was the five-six month training all Marine Officers completed that would provide us with the fundamentals of combat leadership, marksmanship, teamwork, infantry training, and objectively learning the object of war. Marine Corps strategy, in general, was taught with emphasis on leadership and the coordination with units of the United States Navy, of which the Marine Corps is a part. Exposure to gas warfare was provided up close and personal in the form of tear gas ambushes. Reconnaissance training, aircraft coordination, artillery support, and situational infantry leadership were emphasized. Every Marine Officer was expected to be able to command a Marine infantry unit after the completion of the training, even though most officers would never do so. Marine Corps history and acceptable social graces and courtesies were instilled in us in an effort to train each of us to be an "officer and a gentleman."

After Basic School, everyone was given the opportunity to choose which Military Occupational Specialty, or MOS, they would like to pursue. However, there were no guarantees, and the MOS for every Marine Officer would be determined by "the needs of the Corps." Some of the new Second Lieutenants were sent to infantry units with a 0302 MOS because of the ever-increasing need for infantry officers in South Viet Nam...and I was one of them.

When I arrived in South Viet Nam, it became painfully clear how high the casualty rate was among Officers, especially Lieutenants, as was also the case in previous wars. The bottom line was that the Marine Corps needed Officers and they needed them quickly. Minimal training was all that we received compared to the excellent training of Officers today. I can only wonder how our performance as Officers would have been enhanced had we been given more training.

At the end of Officer Candidate Training, I became aware that the Marine Corps had both a basketball team and a football team at the Quantico Schools Base, and that they were used for recruiting. Personnel, both Enlisted and Officers, were allowed to try out for the teams. When I returned to Quantico to begin Basic School in early January 1966, I went to the tryouts and was selected to play basketball. The problem was that I was just entering Basic School, and up until that time they had few, if any, Officers who were in Basic School who were allowed to play on one of the teams. The team practices were in the evenings because everyone on the team had other regular duties. My "bottom line" was that I would leave Basic School at the end of each day and go to basketball practice in a large aircraft hangar. Most games were scheduled for weekends. We traveled almost exclusively by bus, which often led to our arrival back at the Base in the early morning hours. Our travel was mostly to colleges and universities on the East Coast, but the team did fly to Findlay College in Ohio every year on an old C-46 "tail dragger" for a game. Recruiting in the heartland of America was valued, even though most Officer Candidates and Enlisted troops came from the eastern seaboard of our Country.

Games at the various colleges and universities gave us ample opportunity to see the anti-war protestors in real time, and we were not asked to sign autographs. My training was always to come first, and when it came time for the All-Marine Basketball Playoffs, I was not allowed to participate because I would miss too much training. My class standing in Basic School was 42nd out of 470 Marine Second Lieutenants. My Fitness Report for the Basic School included comments that "This Officer participated in extracurricular activities as a member of the varsity basketball team for three months during the reporting period." In addition, it said that, "This six-foot Marine Officer should be considered for future assignment to instructional duty with the Basic School and ceremonial duties with MB, 8th and I."

A life changing moment occurred when I was notified in June of 1966 that Lieutenant General J. M. Masters, Sr., the Commanding General of Marine Corps Schools, Quantico, would like to have "orders cut" to keep me at Quantico for another basketball season. He had the command authority to do that for a period not to exceed one year. I accepted his offer, and I remained at Quantico for the 1966-67 basketball season.

As happens in the military, personnel were assigned to additional duties in other areas, and a different Marine Officer took over the coaching duties for the team. New player personnel joined the team. While I made some enduring friendships that year, the overall experience was not one for the history books, as far as I was concerned. Being raised in a culture where athletic achievement and expectations were held to a very high standard, I could not accept having alcoholic beverages on the team bus, disrespect shown by some Officer players toward Enlisted players, and an atmosphere of compromise by our Coach.

One of our highlights that year was to play Boston College, or "BC." BC was coached that year by Bob Cousy, the professional basketball legend who played for the Boston Celtics only a few short years before. One of our team members was Jim Meyers, who was an Enlisted Marine that held the rank of Sergeant. Nicknamed "Jungle Jim" and "Myth," I believe Jim was one of the most gifted athletes I had ever known. In addition, he was the first black person I had ever known and/or played with as a teammate. Jim was also my first black friend. We remained great friends until his death. Regretfully, we lost many valuable years of personal contact when many Viet Nam Era veterans tried to forget they were ever in the military, as well as the friendships they had established. It was because of a basketball team reunion effort led by another teammate, Captain Bill Sheridan from Long Island, New York and a former player at Fordham University, that Jim and I were reunited after approximately 35 years of separation. Another black friend, who was also a Sergeant, was Willie Sallis. Willie was the son of a minister and we shared many common values and beliefs.

An example of the racism that was evident in the 1960s happened during a game against Hampton Institute in Norfolk, Virginia, an overwhelmingly black student body in numbers, which I believe is now Hampton University. The basketball arena at the time held approximately 5,000 fans, and that night it was standing room only. During the course of the game, I ended out in a "scrum" with one of the black athletes playing for Hampton. As we began "throwing punches," the players for both teams started onto the court, as you might imagine. What happened next was one of the quickest educations a "country boy" from Wyoming could have ever experienced. Jim was on the floor playing, and Willie was on the bench. The instant Willie stepped onto the court to come to my defense, the crowd began pointing at him and screaming "Uncle Tom, Uncle Tom!" Willie turned around, sat down, and refused to participate for the rest of the game. The last time I saw Willie before the 35-year separation leading up to our basketball reunion, I told him I would come to Pennsylvania and look him up after I returned from South Viet Nam. He said, "No, you won't. You would be dead before you got to within five miles of where I live."

Racism was shocking to me then, but it is still "alive and well" in America at the time of the writing of this book. In my opinion, race relations are worse in some ways now than they were back in the 1960s. Bill Sheridan and I attended Jim Meyer's burial service at Arlington National Cemetery in 2014, and we were the only white people there. In fairness to everyone, we were not made to feel welcome by the others in attendance, but I was glad I was able to honor my friend. After the service, all of us went to Fort Meyer, adjacent to the Cemetery, and had lunch. Unbeknownst to Jim's other friends and relatives, I paid for their meals and left for the airport to return to Wyoming. I hope they appreciated my gesture.

Upon the completion of the Basic School, I was contacted by an acquaintance to ask if I would be interested in being in the White House Fellows Program. When I inquired as to the specifics, they said it was a one-year program which included working in the White House in a training capacity. The itinerary would be an in-depth experience of exposure to the inner workings of the federal government from the perspective of the President of the United States. They also said it would require a nomination from a United States Senator from my state, and that Wyoming U. S. Senator Cliff Hansen was prepared to nominate me if I was willing to pursue it. I said I was, and I went to Senator Hansen's Office in Washington, D.C. to discuss the issues surrounding the program with him. He placed my name in nomination, but it required the approval of the United States Marine Corps. The USMC did not approve the nomination because I had not yet completed my tour in South Viet Nam, and I understood without question that my first obligation was to my Country as a Marine Officer. However, I was extremely honored that Senator Hansen recommended me for such a prestigious honor.

My Officer Candidate and Basic School comrades had been gone a year by the time I finished my temporary assignment to play basketball at Quantico, which included perfunctory duties as a young Second Lieutenant. I actually had a command responsibility of mostly "short timers" who were nearing the end of their enlistments. Because of my faith in people and my confidence that everyone could become a productive citizen in society and a military unit, I was assigned Marines whose conduct created specific problems for the military. My first lesson in reality involved a young Marine who had been in the Marine Corps for two years. He had never been on leave, and his only responsibility had been KP duty. KP was the acronym used for working in food service. I was sure I could help this young Marine.

Through extensive preparation and personal attention, I had him ready, or so I thought, to go on leave for the first time on Labor Day Weekend, 1966. When I arrived at the barracks at 0530 hours on the day of his scheduled departure, my First Sergeant was waiting and told me we needed to go to the Brig, or military jail, on the Base. On the way, he told me that our young Marine that I had spent endless days, weeks, and months trying to help to become successful in the Marine Corps had viciously stabbed and killed another Marine the night before at the Enlisted Club. Supposedly, he and the victim were in a fight over a married woman whose husband was in South Viet Nam.

When I was given the opportunity to remain at Quantico for the additional basketball season, I had sought advice from senior Officers about the wisdom of remaining there for that year. Several cautioned me that resentment existed amongst certain Officers toward athletes in general. Supposedly, those Officers contended that "special privileges," real or imagined, were extended to some of us because of our athletic abilities. They warned me that those sentiments could be reflected in my future Performance and Fitness Evaluations, which were submitted every six months. Their concerns were verified in my Fitness Report from 30 July through 31 December 1966. In that Report, my Commanding Officer rated me as Average in eight of 16 categories, Above Average in five categories, Excellent in two categories, and Outstanding in the category of Loyalty. The CO also indicated the he would **"be glad to have"** me under his command. As the Reviewing Officer of my Fitness Report for that period, Lieutenant General J. M. Masters, Sr., who had issued the orders to have me held at Quantico for the additional basketball season, wrote the following:

> "In my judgment these marks do not reflect the true ability or potential of this young officer, who has demonstrated as a member of the Marine Corps Schools Basketball Team a fine ability to lead. My Observation of him is that he is above average to excellent in performance as a Marine officer and has excellent potential. I would particularly desire to have him in any capacity."

One reason for having Reviewing Officers of Fitness Reports was to prevent what happened to me. The apparent bias of my Commanding Officer was addressed by that Superior Officer in writing. I have always appreciated the General's intervention on my behalf. My Fitness report for 1 January through 11 July 1967 from the team basketball coach included 10 categories in which I was rated Excellent and three categories that were Outstanding. My organizational Commanding Officer for that period rated me as Excellent in six categories and Outstanding in three categories. Both Officers indicated that they would "**particularly desire to have**" me under their command. My Marksmanship Qualifications with the Rifle and Pistol were both recorded as Expert.

In the late spring of 1967, I was assigned the duty of helping organize and carry out the operation of the Quantico Relays held on Marine Base, Quantico, Virginia. The Quantico Relays was an invitational track and field event that attracted many of the leading track and field athletes in the nation. It was an outstanding experience that produced memories of a lifetime for me. In late June, I returned to Wyoming to enjoy some leave with my family in preparation for my departure for duty in South Viet Nam.

My orders were to report to the Treasure Island military facility near San Francisco in early July for processing and transfer to Camp Hansen in Okinawa. I took a connecting flight from Cheyenne, Wyoming, through Denver, to San Francisco. There were no "well-wishers," waving flags, bands, military personnel, or dignitaries to bid me farewell as I boarded my flight. I was not part of a Marine Unit being deployed, just leaving on a deployment as an individual, which was not uncommon for Marine Officers during the Viet Nam War. To be fair, few people in Wyoming at that time had more than a geographical idea where South Viet Nam was, and very few people from Wyoming had been sent there as a member of the United States Military. At that time in the summer of 1967, my dad was just getting ready to begin harvesting his wheat crop. As I boarded the flight, I had confidence that he would be able to get his crop in with the help of my brother's two sons, Billy and Rick Chamberlain. They had come to La Grange to help him, since I was leaving for my overseas tour of duty.

I was proud to serve my Country, and I still hope most of my family was proud of my service. However, the carnage of war changes everyone forever. The harsh reality is that the families and loved ones of the combatants also had to suffer the uncertainty of a separation involving unknown dangers and very limited communication. Living with what is left of a person mentally and physically in the aftermath of a combat experience very often leads to broken relationships, divorce, and even more tragically, suicide.

The despicable way Viet Nam Veterans were, have been, and some still are treated by many Americans is a travesty that will hopefully never be forgotten and never be allowed to happen again in our great Country. The carnage of broken dreams, hearts, and minds of Veterans who served in the Viet Nam theatre of operations is immeasurable. It is beyond belief that, while those of us who are Viet Nam Veterans had to stay in the shadows for 30-plus years and often deny we had served there, we must now watch some of those who refused to serve their Country lie about what they did when our Country called, especially when they claim they are Viet Nam Veterans when they are not.

Destination: Okinawa

ACTIVE DUTY MILITARY PERSONNEL had to travel in uniform in those years, and on the flight from Denver to San Francisco, a flight attendant quietly approached me in my Economy Class seat and asked if I was on my way to South Viet Nam. When I confirmed that I was, she said there was an empty seat in First Class and that I was welcome to have it. It was the one and only time I have ever flown First Class, and I will be forever grateful for her act of kindness and consideration that day, despite the apparent frowns of disapproval displayed by some other passengers.

My arrival at Treasure Island was uneventful, and my stay was approximately 48 hours. During the processing, I was handed a M-16 rifle with one clip, or 20-round magazine, of ammunition. I was told to go to the small firing range and fire the weapon to familiarize myself with it. It was the first time I had seen a M-16, and little did I know how many countless Marines would be killed or wounded because of the unreliability of that piece of armament.

My next flight was a commercial charter flight. Those on board all appeared to be military personnel of some kind, but I did not see any other Marines. It left San Francisco and it was a direct flight to Tokyo, Japan, where we arrived in the late evening. We were given sleeping quarters in a hotel close to the Tokyo Airport with very limited facilities. Most of us had to sleep on the floor in the rooms assigned to us because there were six of us in each room, and the rooms were designed to accommodate two people. After about five hours, we boarded another commercial charter flight to Okinawa.

Camp Hansen, Okinawa, was the processing center for all Marines going "in-country." We received our final shots and inoculations required of all personnel. They included tetanus, hepatitis, malaria, and typhoid fever, just to name a few. Because of the large number of people processed each day, they used a relatively new air-actuated shot mechanism. It was shaped like a large pistol and it did not require the use of a needle to inject the vaccine materials. Instead, it had a round, flat head on the end of the "barrel" about the size of a 50-cent piece with a hole in the middle of it. The Corpsmen would press it firmly against our arms and actuate the trigger. The liquid containing the product was forced into your skin and muscle with a very quick burst of air. The contention was that it was more sanitary than needles, and it avoided the problem of supplying needles in the quantity that would have been needed.

One of the problems with the procedure was that if the round metal ring was not evenly pressed against your skin, or you flinched or moved, it would cause the injection to go in at an angle. If that happened, the injection would usually tear the skin and muscle under the skin and then exit the skin somewhere near the injection at an angle. The resulting wound would sometimes require stitches to close the gaping injury. We stood in unbelievably long lines and when it was your turn, you stepped between two lines of medical personnel who administered shots from both sides at the same time. In seven steps forward, we were given fourteen shots in a period of approximately one minute.

At the end of that line we were told to "Drop your drawers and bend over." As soon as the skin was drawn tight on our buttocks, one medical personnel would slap us on the buttocks with a paddle shaped object and another Corpsman would quickly inject Gamma Globulin into our "butts" with a needle. If the procedure worked as expected, the sting from the slap helped minimize the pain from the shot. The shot in the buttocks could be felt in our posterior for the next 36–48 hours and gave us the sensation of sitting on an egg every time we sat down.

The dental examination was basically, "Open your mouth," followed by a cotton swab drenched with liquid fluoride being shoved into our mouths. The fluoride was smeared quickly on our teeth, and that was the last dental care most of us would receive for the next 13 months.

Prostitutes were available in the areas around Camp Hansen where we could go on "liberty call," and some of our personnel availed themselves of their services. We were told to make sure a prostitute had an updated shot card indicating that she had been treated for possible venereal disease before we engaged in sexual relations with them. With my background and spiritual beliefs, I could not believe that people actually engaged in that sort of conduct. I remember wondering how disease was actually controlled in those kinds of conditions, and after I took my command in-country, I learned that, in fact, it was not controlled. I later used inventive measures to deal with the problem of venereal disease when new troops joined my Company in South Viet Nam. However, in the atmosphere of political correctness in our Country today, it would probably have been considered illegal. I will return to that subject in more detail later.

We were issued sea bags to store our personal items in. We were told they would be secure, and we could pick them up on our way back to "the world." It was suggested that we not take any jewelry of any kind in-country, including rings and watches, nor anything else that would reflect light or bring attention to ourselves. That warning proved to be very important once we arrived in South Viet Nam

In-country

A NORTHWEST ORIENT AIRLINER was our mode of transportation from Kadena Air Force Base in Okinawa to DaNang, South Viet Nam. It was a Boeing 707 charter flight that had every seat in it that could possibly be placed in a plane that size. The passenger manifest again was a mix of what seemed to be personnel from all of the various branches of the military. It was a night flight that took approximately four hours. As we neared South Viet Nam, word was passed that earlier in the day someone was wounded by ground fire that came through the bottom of the plane as it was approaching the DaNang Air Base, and that was one of the reasons we were being flown in at night. As soon as the plane touched down, the landing lights were turned off and the plane taxied in darkness. The two flight attendants who had given us water to drink on the flight moved quickly to the rear of the plane.

When the plane stopped moving, the door just behind the cockpit opened quickly and we were told to get off and down the portable stairs as quickly as possible. Luggage was not a problem because all we had were our limited number of military papers and the clothes we had on our backs. The engines on the 707 were left at idle and we hurried down the stairs where we jumped in the backs of military "six-by" three-axle trucks for a very short ride across a couple of runways where we were told to jump into some sandbag revetments. They were enclosures made of sandbags stacked about 3-4 feet high.

As we jumped down behind the sandbags, the Northwest Orient airliner that brought us taxied and turned onto the runway right next to us and, with an ear shattering burst of full power, it was quickly airborne into the darkness. Since I had left Cheyenne, Wyoming, I still had not seen anyone whom I ever saw or knew before, and that night did not seem to be the time to make any new acquaintances. As I laid there in the darkness, I wondered how I could have gotten myself into that situation. I was supposedly in a war zone, fighting for my Country, and I did not even have a weapon.

As that first night wore on, we experienced the ear piercing and sensory shattering sound of jet engines powering fighters and bombers of various sizes on their missions throughout the Viet Nam Theatre. The noise was so overwhelming that several people behind our wall of sandbags became nauseous. Unknowingly, I was seeing aircraft up-close and personal that would save the lives of my troops and me and help us accomplish our missions on numerous occasions in the future. I had no idea the very significant part those "airborne giants" would play in the ground war in Viet Nam that I was about to become a part of. I longed again for the chance to fly one of those outstanding machines. However, the first light of day was a welcome sight...but it seemed like it would never come.

2nd Battalion, 7th Marine Regiment, 1st Marine Division

My duty assignment in South Viet Nam was to the 2nd Battalion, 7th Marines, or "2/7" for short. It was an Infantry Battalion with all of the supporting units that were included under the Marine Corps operational organization at that time. The United States Army infantry units may have been organized somewhat differently, because the missions of the Army and Marine Corps usually differed somewhat. However, in the Viet Nam War, the Marine Corps took on a more traditional Army role of being a stationary force in a part of South Viet Nam where we were responsible for all operations in a designated area. The Marine Corps area of operations was mostly limited to the I Corps area of South Viet Nam, which was the area from the Demilitarized Zone next to North Viet Nam on the north, to south of Chu Lai on the south, and to the country of Laos on the west.

The 2/7 Battalion Headquarters were located west of DaNang, and one of the areas the Battalion was responsible for was the "Rocket Belt." The Rocket Belt was an area drawn in a half-moon shape on a map from where the North Vietnamese 122-millimeter and 140-millimeter rockets could be fired and reach the DaNang Military Facility. Because the 122 mm were less powerful and had a shorter range, the term "belt" was used to cover the minimum range for the 122 mm to the maximum range of the 140 mm.

These rockets and/or rocket components were small enough to be transported by personnel and pack elephants, and because they could be assembled rather quickly, they were able to be launched from hand-dug emplacements. The ground would be sloped at an estimated angle to determine the trajectory of the rocket, and a simple weighted object of anything hanging on a string was used for aiming purposes. Because the rockets could be prepared and launched in a short period of time, and the persons firing them could be gone quickly, they often avoided any return fire from our artillery or aircraft or pursuing ground units like ours. Their accuracy was amazing, especially considering the modern technology we had at the time compared to their rudimentary technological capabilities. Obviously, one of our main ongoing missions was to prevent our enemies from reaching the Rocket Belt with those armaments and, if they did, not to allow them the time to launch them. In the jungle and other terrain, that mission was very difficult to accomplish.

Learning the Actual Military Tactics and Operations of a War

WHILE I MAY HAVE THOUGHT I knew what General George Patton meant when he was speaking to the West Point Cadets in his speech I referenced earlier, the actual combat experience in South Viet Nam was considerably more life changing than I ever imagined it would be. The legendary actor John Wayne has been quoted as saying, "Courage is being scared to death...but saddling up anyway."

Fear is a reality in the combat zone of a war. The immediate task at hand for everyone who finds themselves there is the management of that fear. If the fear becomes "all-consuming," the person who is affected may become a danger to themselves and others. Fear has to become a reality that there are significant dangers that must be addressed, but in a manner that reflects the person's training. In the Marine Corps, Marines are taught to instinctively react, follow orders quickly, and to rarely consider other possible options after orders are given.

It normally took 18 months in grade as a Second Lieutenant in the Marine Corps before an Officer could be promoted to First Lieutenant. Because of the extra year I had been at Quantico playing basketball, I was technically promoted to First Lieutenant enroute to South Viet Nam. In an infantry unit, the normal command duty for a Second Lieutenant was that of a Platoon Commander of approximately 35 men. When I arrived at my new duty station with 2/7, my previous command experience at Quantico had been very limited, and none of it was as a Platoon Commander in what was referred to as "the field." While I was stationed at Quantico, I had occasionally acted as a Trial and Defense Counsel in

Court Martial cases when I was performing that temporary additional duty, but that was entirely different than what I was assigned to do now.

The normal command responsibility for a Marine First Lieutenant in an infantry unit during that era was that of being an Executive Officer, or second in command, of a Company, which could have included as many as 220 men, depending on the exact organization. That being the case, the Battalion Commander, Lieutenant Colonel Wilson, decided I needed to spend more time with the Battalion Headquarters than the average 1st Lieutenant.

With that in mind, he assigned me to be the Assistant Battalion S-3, or S-3A. If possible, most newly assigned First Lieutenants and Captains worked in the Battalion Fire Direction Center for a period of time upon their arrival to build an understanding of the intricacies of commanding an Infantry Company. The training was designed to include the knowledge required to perform basic infantry unit communications; coordinate emergency medical evacuations; access artillery fire and naval gunfire support; communicate with airborne forward control observers and coordinators, fixed-wing aircraft, and helicopters; establish "save-a-plane" notifications; and obtain supply and resupply distribution. Learning the time it took to perform these various functions was absolutely essential, especially in an emergency situation that required abbreviated procedures that could make the difference between the life and death of my men. Lt. Col. Wilson was rotating back to the world soon after I arrived, so my experience with him was limited.

Shortly after I arrived, we went on a Battalion-sized operation. An Infantry Battalion was comprised of four rifle companies with support units and could easily have been 700 Marines. During my time with 2/7, I did not see many of those. There was very little enemy contact, but a memorable incident happened on a "pitch black" night when everyone was in their foxholes or holes. Lt. Col. Wilson suddenly let out a muffled shout and we could hear him scrambling away from his hole. It sent a tremendous "spook" through everyone within hearing distance. He soon quietly explained that he had sensed movement in front of him in the darkness, and he thought it could have been a snake. We checked the area with our filtered flashlights and found a cobra snake in his hole. We surmised that the snake had raised itself up and was swaying back and forth as those snakes do to apparently hypnotize their prey. Fortunately, Lt. Col. Wilson somehow noticed or sensed the motion. He said he slowly clenched his fists into the dirt on the sides of his hole, and all in one motion he shouted, threw the dirt, and jumped out of the hole. The snake apparently struck at him and fell into the hole. One thing about the jungles...you could see many of the animals you may have seen in a zoo, only these were in the wild and often "up close and personal."

Lt. Col. Wilson was replaced by Lt. Col. John Love, the finest Marine Commander I ever had the privilege to serve under. He graduated from the United States Naval Academy in 1951. After being commissioned as a Marine Second Lieutenant, he served in the Korean War in the Spring and Summer Campaigns in 1952. In 1967, he received orders to report for duty in South Viet Nam. One of my greatest regrets is that I will never be able to share my writings with him. When I tried to contact him about my memoirs, I discovered that he passed away in 2010. I will simply refer to him as "Col. Love" hereafter. He was a "hard charger" whose leadership goal was always to be of assistance to his Company Commanders, and he was. I will be eternally grateful to him.

Soon after Col. Love took command of 2/7, he told us he wanted to go on a reconnaissance mission so he could actually see the problems his troops were dealing with in the jungles. I admired his foresight for wanting to do so, but I wondered about the wisdom of his plans for that particular Operation, even though I was new. Dealing with people in the Battalion Headquarters was always difficult because they often "had not been there or done that," and he was trying to avoid that. However, my concern in that particular case was the problem of providing the tactical security for a small unit of 30-35 troops, and for his safety because of the danger to him personally in a limited Operation of that size. I could readily imagine the possibilities of him being killed or captured on jungle trails where vision was often limited to a few feet or less.

On the day of the mission, we were dropped off by Six-by trucks near a banana grove where a trail went into the mountains northwest of DaNang. That trail was known to us as a route of ingress and egress into and out of the DaNang area by the North Vietnamese and Viet Cong. It was hot, and I thought we were moving too fast for the temperatures and conditions along the trail according to my training. However, I was not about to try to correct an Officer of his rank. Even before the sides of the trail began to close in and restrict vision and movement, I could see the Marines in the lead were beginning to show the effects of the heat, and they did not seem to be as cautious as I thought we should be. However, who was I to be "second guessing" a Lieutenant Colonel? I had not done this before.

About a mile or two up the trail, we were ambushed and hit with automatic weapons fire. By this time, the jungle was so thick along the trail that you could only see the trail if you were looking down at the ground, and the foliage was so thick we could not even get down toward the ground to try to keep from being shot. Fortunately, the thickness of the jungle also helped limit the range of the rounds, and Col. Love was not hit. However, the Marine on point was killed and the two troops behind him were wounded.

According to my previous training, it was a classic ambush on a trail, and I learned that day it was for real. I think it brought the realization to me that we were there to kill people...and do it before they killed us. It was a long way from playing "Cowboys and Indians" in La Grange, Wyoming, with my childhood friends. After assessing the situation, Col. Love decided to move on up the trail. He left a radio operator and one other Marine with me and told me to get the dead and wounded evacuated. I will admit I was not sure how we were going to do it, but I give credit to my rural, agricultural background for having the presence of mind to figure out a way to get it done.

It seemed like an eternity before the medevac helicopter arrived.

Catastrophic mechanical problems had developed with the large, twin-rotor Marine CH-46 helicopters shortly after my arrival in-country, and the main support aircraft for all Marine land operations were all grounded until the cause of the mechanical failures could be remedied. For some reason, the bearings on the synchronization shaft of the twin rotors appeared to "seize up," causing the shaft to break. The loss of synchronization of the two large rotor blades would cause the helicopter to break apart. It could happen at any time, and if it happened when the chopper was airborne, the results were always catastrophic. Without the main workhorse of the Marine Air Wing, we were left with the CH-34 Korean War Era single-rotor helicopter that was powered by a 16-cylinder gasoline-fueled radial engine that had very limited power and lifting capacity. To complicate matters, many of the CH-34s did not have jungle penetrators, which was a winch driven device that could be used to lift and extract wounded and dead Marines upward to the helicopters hovering above the jungles. They were necessary because there were many areas where choppers could not land.

The jungle penetrator, when folded up, looks like a cylinder that is pointed on one end and is hooked to a cable on the other. The cable is attached to a winch on the helicopter. By pushing a button on the penetrator, four arms fold outward from the pointed end and the arm was placed between the legs of the person to be evacuated. The casualties were placed on the arms in a sitting position, and there was a strap similar to a seat belt located by each of the penetrator arms. Once the person was sitting on the arm, the seatbelt was placed around the abdomen of the person and pulled tight to hold them upright in an attempt to try to keep their head from tipping back. If their heads tipped back, the tree limbs could damage their face and eyes as they were being lifted up through the jungle canopy. In some places, there were three layers of canopy that could extend as high as 250-300 feet.

As the chopper approached, I told the pilot we had only been able to find one small area where we could possibly complete the evacuation. When he saw the area, he told me it would not work because he did not have a jungle penetrator. In a relatively quick discussion, the pilot said he could see an area a short distance away where he would try to land on top of a small tree to see if he could crush it toward the ground. We agreed it might enable us to lift the casualties high enough so his crew could get ahold of them. Everything worked as planned until we could not lift the casualties quite high enough for the crew to reach them. What seemed like an eternity passed before the crew was able to get ahold of our dead and wounded troops, and all the time the pilot was balancing the chopper on top of the tree and other jungle brush. The continuous "prop wash" under the rotor blades of the chopper above us made it almost impossible to move, let alone lift the bodies into the air. However, our persistence and teamwork with the chopper crew finally paid off, and the chopper lifted off of the tree it had been balanced on by the skillful and courageous pilot.

That experience was one of many involving helicopter crews, who I believe were undoubtedly some of the bravest personnel I ever served with. As darkness closed in, the three of us set up in what I thought was a defensible position in the jungle close enough to the trail to be able to possibly ambush an enemy unit that might come along the trail during the night, and also be able to hear Col. Love and the rest of his "reconnaissance team" when they returned the next day.

During that night, I mentally reviewed what I had seen and done that day. The dead Marine we medevacked was black, and I marveled to myself at how red his blood was. Keep in mind that I had a college education. How could I have thought that the color of his blood would be anything else? It was a moment of intellectual confrontation and stark realization that I described many times to my students later when I was teaching in public high schools, and when we would discuss racial discrimination. It always seemed pertinent for me, in each appropriate discussion, to remind my students that, as humans, we all have the same blood. Only the color of our skin may be different.

Col. Love spent extra time with me during the next few weeks in preparation for my becoming the Executive Officer of one of the

2/7 Companies. One of the things I appreciated about him was that I always knew exactly what he was thinking concerning any subject or conversation. I remember how amused he was when we had a rocket attack on the Battalion Headquarters. I ran to the Fire Direction Center bunker to be of help and continue my learning. Col. Love was already there. There were railroad ties used for steps. I tripped on the last one and fell just as I tried to get into the bunker. As I fell, a burst of enemy AK-47 rifle fire impacted just above my head. If I had not fallen, this book would probably never have been written. As you might imagine, I did not think the incident was nearly as humorous as he did.

In a letter from the world, I learned that a "shirttail" relative of mine, who was a United States Army helicopter pilot, was also in South Viet Nam. Army Captain Max Beebe and I had attended John Brown University together, and we had married women who were first cousins. I had grown up with both women in our small Wyoming town. Since I was near the end of my orientation training at 2/7 Headquarters, I asked Col. Love if I could try to get to Captain Beebe's unit for a quick visit. He agreed, and I inquired of an Army unit near DaNang about the location of Max's unit. I was able to get a ride to his squadron base, and when he returned from a mission that afternoon, we had a couple of meaningful hours together before I returned to 2/7 later that day.

"Go Get 'em, Tiger"

By the first of October 1967, it had been almost exactly two years since I entered OCS. I believed I was ready to be an Executive Officer, or XO, of a Rifle Company, and Col. Love agreed. 2/7 consisted of four rifle or infantry companies, Echo, Foxtrot, Golf, and Hotel. He made the arrangements for me to go to Hotel Company on Thursday of the first week of October to become the XO. On Monday of that week, Captain Lyle Johnson had assumed the command of Echo Company because the previous Company Commander was ending his tour of duty and was rotating back to the world. Two days later, on Wednesday, Captain Johnson was killed in an ambush on Highway One north of DaNang.

Because of the high casualty rate amongst Company Grade Officers, meaning Captains, First Lieutenants, and Second Lieutenants in South Viet Nam, there were very few Officers who were not already fulfilling command responsibilities throughout the various organizational elements in Infantry Battalions. In our Battalion, I was the only one available with a rank above Second Lieutenant. Col. Love called me into his "hooch" and told me his dilemma. He chewed on his cigar for a few seconds and then said, "I believe you can handle Echo Company. Go get 'em, Tiger." I was stunned, but I will never forget the pride that I felt because he had that depth of confidence in me.

I had little time to think about what I should do as a First Lieutenant with 3-4 months in grade. It meant after only being a First Lieutenant that short period of time, I was being given the command of an Infantry Company, a position that was supposed to be filled by someone with the rank of Captain. The fact that I had no experience as a Platoon Commander and had never been an Executive Officer of a Company made it seem almost unbelievable. I have never forgotten what he did for me that day, but I was to learn later that my entire future in the United States Marine Corps and beyond would also be in Col. Love's hands.

As I recall, a Rifle Company at that time was supposed to consist of 220 Marines. However, I do not ever remember having over approximately 180 troops at any one time in the field. Casualties, rotations back to the world, time needed for replacements, and illnesses were primarily the reasons for the reduced numbers. Echo Company had outstanding NCOs (Noncommissioned Officers), and Staff NCOs (Staff Noncommissioned Officers). These ranks, in addition to Private, Private First Class, and Lance Corporals, were referred to as being in the Enlisted Classification of Marines. Marine Officers like Second and First Lieutenants and Captains were referred to as being in the Officer Classification.

Unbeknownst to me at the time, when the word spread through Echo Company that the new Company Commander did not have any combat experience, there was obvious concern amongst my men. Their concern and understandable anxiety are described in my then new Company Radio Operator "Cathi's" book *Shaped by the Shadow of War* (page 375):

> Lieutenant Chamberlain had not served as a platoon commander in Viet Nam, thus, his first command was at the company level. Battalion had him lined up to be X.O. of another company, so that he could follow the same learning path that Lieutenant Johnson was on. But the sudden loss of Echo Company's C.O. forced the Battalion Commander to install the young lieutenant earlier than planned. To make matters even more irregular, the reason Doug Chamberlain didn't serve as a platoon commander was because he was busy playing on the Marine Corps basketball team. Word about this spread rapidly and more than one Marine bemoaned our fate to have a commander who had never commanded.

As I would have in any position of leadership in private business, I sat down with all of the NCO and Staff NCO Marines in my new command for a frank discussion as soon as I arrived. The biggest problem I confessed to them was, in my opinion, my lack of experience. I discussed my expectations with them that were centered around my idea that all the members of our Company must be part of the effort and show initiative, not just be waiting to be told to do something. I stressed that I thought all of us could play a key part in gathering intelligence because every Marine should be the "eyes and ears" of the Company. I also told them I wanted to develop a "grapevine" so that information could get to me quickly if someone had a concern but did not want to be personally identified as having said something. It was obvious to me that having the correct information at my disposal at the right time could save lives. My expectation was also that every leader from the Fire Team, to the Squad, to the Platoon Sergeant, and the Company Gunnery Sergeant should be trained to take another position of leadership above their own in case the situation demanded it. It was a "next-man-up" approach as is sometimes referred to in professional sports. It even included becoming Platoon Commanders, which we obviously were already doing, since one Second Lieutenant was the only other Officer in the Company other than me at the time. I should have had four Lieutenants, instead of just one. Because my Company operated on its own most of the time, some of our enlisted leadership had to remain at the Battalion Headquarters to handle the Company administrative responsibilities. The Marine who was acting as the Company Gunnery Sergeant, or "Gunny," was in fact a Staff Sergeant and, who like me, had responsibilities above his rank. Because of the shortage of Lieutenants, I sometimes had Sergeants acting as Platoon Commanders. As I noted earlier, when I took command of Echo Company, without reservation, I took command of some of the finest men, who happened to be Marines, that I have ever known.

According to some of my Staff NCOs and NCOs, Echo Company had existing problems that appeared to be the result of a lack of supervision over a previous period of time. Regardless of those allegations, I began the task of learning what problems needed to be solved the day I took command. The Lieutenant I had was Second Lieutenant Leonidas Caesar Stair, III, an Ivy League Graduate of Yale University. His acquaintance was the beginning of a friendship that has endured to the date of this writing. I instantly dubbed him as "Sid," which he still answers to today.

Another Marine stepped forward in those first days named Sergeant John "Jack" Johann. Jack had been in the Marine Corps previously, but he had left the Corps to enter law enforcement as a police officer in Connecticut. With the commencement of military actions in South Viet Nam, he reenlisted in the Corps, even though the USMC reduced him in rank as a condition for allowing him to re-enlist. Without reservation, he was and is the most qualified Marine of his rank that I have ever known. His actions as related later in this book have emblazoned him in an honored place in Marine Corps history, in my opinion.

Corporal Donald Catherall, from Texas, enlisted in the USMC to serve his Country. He had been the Radio Operator for Captain Johnson before he was killed and "Cathi" became my official Company Radio Operator. He was the epitome of an outstanding Marine, and his assistance to me was beyond adequate description. Sid, Jack, and Cathi became my closest confidants during my tenure in South Viet Nam, and they still fill that position in my life today, in several ways. Cathi described the beginning of our relationship between me as an Officer and him as an Enlisted Marine from the following perspective in his book on pages 374-375:

> *Normally, company command is a captain's billet, but the reality was that first lieutenants often commanded companies in Vietnam, because there weren't enough captains to go around. Assuming command of a rifle company in a combat zone is pretty much a guaranteed way to make captain, however, and Lieutenant Chamberlain soon became Captain Chamberlain. Lieutenant Johnson was posthumously promoted and buried as a captain, which pleased me.*

He elaborated on page 375-376:

> Most of the Marines in a rifle company know the company commander from a distance. When I was in a platoon, I never exchanged a single word with the company commander, and only very rarely did I speak to my platoon commander. It was amazing how much that had changed in only a few months. I would be at this new C.O.'s right hand whenever anything was going on. More than anyone in the company, I would see exactly how he performed in the field.
>
> I'd been through a similar process with Lieutenant Johnson, but it felt different this time. I was more confident after all I'd learned with Lieutenant Johnson. I felt I had something to offer this new commander. I saw him less as father figure and more as a guy needing all the help he could get to do a very tough, and very important, job.
>
> Today, I look back at that transition as the biggest turning point in my life.

The Echo Company mission on the day I assumed command was to guard the ESSO fuel facility along Highway One north of DaNang where ships in the DaNang Harbor unloaded motor fuel into storage for use in civilian vehicles. The mission included providing protection for Navy Seabee Units who repaired the damage to the roads and bridges on an ongoing basis because of daily attacks by Viet Cong and North Vietnamese Army Units. The 26 bridges were spaced randomly throughout 19 miles of road from the ESSO facility up over Hai Van Pass to the north boundary of the Echo Company Tactical Area of Responsibility, or TAOR, which was near the town of Phu Bai.

Life in a combat zone does not fit the "stateside" expectations as far as military protocol. Saluting someone could get one or both of you killed by a sniper, because the recognizable motion indicated one of you must be a higher rank than the other or a specialist of some kind, so we always avoided doing it. We ignored the protocol of having Corpsmen Medical Personnel, or Corpsmen, unarmed and made every effort to make them look like combat troops because they were one of the first targets in any contact with enemy forces. There was no, "Yes, Sir," or "No, Sir," and everyone knew who everyone else was...or they had better know. The names and nicknames were numerous, and there was the name "Skipper"...that was me.

Once in a while, we would have two Marines named "John Smith," for example, and it sometimes made for excitement. I received a radio message that "John Smith's" wife had delivered a baby boy back in the world, and the mother and baby were doing fine. We found John Smith and began to congratulate him and pound him on the back. We were as proud as we thought he would be. John Smith was very quiet, and even shocked by such a loud celebration. When the cheering and shouting finally stopped, he quietly said to me that he "just couldn't believe it." And then he said, "I'm not even married." Needless to say, we had the wrong John Smith, so when we found the actual "proud Daddy," we "hollered and howled" even louder than at the first mistaken announcement!

The first night with my new Company was memorable, but disheartening. My Company Gunny came to me about dark and described an overall situation where, if it were true, the security of our Company and the ESSO plant were in jeopardy. I had no reason to doubt him, but I had only met him a few hours before and it was hard to imagine that the situation could be so dire. I told him we would check our holes together. Sadly, the situation was just as he had described it.

At about 2200 hours, we approached a hole and found both Marines asleep. We removed their weapons, ammunition, flares, and other munitions. Neither of them even stirred. I nudged one with the toe of my boot, and he was startled when he awoke. I asked who was supposed to be on watch and he indicated that the other Marine was. About that time, the other Marine awoke rather casually. I told them both they were risking the lives of every Marine in the Company. The Gunny told the Marine who was supposed to be asleep that he should take care of the situation. Suddenly, the Marine who was supposed to have been asleep punched the other Marine who was supposed to have been on guard, and a fight ensued in the hole. However, the Marine who was supposed to have been on guard began to win the "rumble." Gunny grabbed one of the "combatants" and pulled him out of the hole. He administered what was referred to as a "thumping," or "ass kicking," and then dragged the other Marine out of the hole and did the same thing to him. Gunny explained to each of the young Marines as he "worked them over" that one was "getting the shit kicked out of him" for being asleep on post, and the other "got his treatment" for not being able to "kick the other S.O.B.'s ass." Needless to say, both Marines were given discipline that was "not by the book," but which I am sure was remembered for the rest of their Marine Corps enlistment.

At the end of the thumping, we told the Marine who was supposed to have been awake to stand 20 feet in front of the hole outside of our Company perimeter without his weapon. We gave both of their weapons and ammunition to the other Marine, who was supposed to have been asleep. We told him he could let the Marine who was sleeping return to their hole when he thought the offender had learned his lesson. When the Gunny and I returned at about 2350 hours, the Marine in the hole still had the other Marine standing 20 feet in front of the foxhole unarmed, and he had his rifle aimed at the offender.

As we moved around the perimeter, we found an area where only every other hole had a Marine awake and on guard, which allowed four Marines out of every three holes to be asleep and created large openings in our defensive perimeter. Gunny again engaged the offenders, but this time a tooth was broken, as well as a lip split. The Marine needing dental work said he was going to report Gunny for maltreatment, but by morning, I think the other troops in the Company had convinced him that all of their safety was worth the "slackers getting what was coming to them." The report I received later after my trooper was sent back for tooth repair was that he said he "fell and hit his tooth on a rock." In those years, military justice could provide for capital punishment for sleeping on post in a combat zone, depending on the situation.

Gunny then took me to an area inside of our perimeter wire where a culvert was buried in an embankment. He shined a light into the culvert illuminating two Vietnamese females sitting there who began to giggle nervously. He had explained to me that for a considerable period of time, local "prostitutes" from a nearby village had been crawling through the wire to wait for our troops in the culvert. The security risk for my Company was unimaginable. We pulled them from the culvert and took them to where our perimeter crossed Highway One. Gunny opened the wire, and I tried to make them understand they would be considered "VC" if they returned. We shouted, "De, De!" and I fired a few rounds over their heads as they ran down the road to make sure they understood what was going to happen to them if they ever returned. That seemed to solve the problem. It was a night of little sleep, and I had a grim realization that there was considerable work to do to return Echo Company back to a position of being what I considered to be a good combat unit.

Transportation on Highway One was difficult at best, even on foot. The vehicle Echo Company had assigned to it was what was referred to as a Personnel Carrier, or PC. It was a little larger than a modern day one-ton pickup like a Ford F350 with a long box and single rear wheels. However, the PC had been used by Captain Johnson and his Radio Operator to go over Hai Van Pass the day they were ambushed and killed. The rocket the enemy forces used in the ambush destroyed the vehicle.

The next best vehicle we were given to use was a "Mighty Mite." It was a very small "piece of scrap metal" made by American Motors Corporation, as I recall. The "worthless piece of junk" had an air-cooled engine that I think had four cylinders...if they were all firing. It had a manual transmission, a canvas top, two seats, and a very small cargo area behind the seats. It was built to be of such light weight that it could be dropped from an aircraft with the use of a parachute. It was so poorly designed and constructed that the wheels were attached to the spindles/axles by a single, small bolt. When we were going up Highway One a few days after I took command of my Company, the bolt on the front wheel on passenger side broke allowing the wheel to come off. The wheel rolled over the cliff on the edge of the road and down into the canyon below.

Since I had learned in my rural Wyoming upbringing that "Necessity is the motherhood of invention," I had confidence my driver and I could find a short-term solution. We decided that I should brace my feet on the little rear bumper on the driver's side and hold onto the fender and cargo area. We thought that if I let my body "hang out over the road," it might be enough weight to cause the opposite front corner of the vehicle where the wheel was missing to clear the ground. Luckily it worked, and my driver was able to get us back down the narrow, winding road. My driver always maintained that it was his driving skills and the dead weight of my "skinny butt" that made it possible to get to the bottom of the mountain. I am sure the gooks really had a laugh if they saw that maneuver.

Shortly after I took command of my new Company, I went up Highway One past the place where Capt. Johnson had been killed. I wanted to review the defensive positions my troops had at various bridges on the way up and over Hai Van Pass. In the pending darkness, I checked the defensive positions at one of the largest bridges where Sid had deployed part of his Platoon. The bridge was located next to a large area of boulders dislodged during the original road and bridge construction. The boulders had been piled on the side of the road opposite the mountain. Behind the boulder area was a very steep cliff where rock chips, dirt, sand, gravel, pieces of trees, brush and other refuse had been pushed into the valley below during the bridge repairs needed after the gooks had tried to destroy it. It appeared that it would be very difficult for anyone to climb up through all of that loose material without making noticeable noise.

The boulders had enough space between them that they provided a very defensible position to protect against any fire coming from the highway and mountain slope above it. There were excellent fields of fire toward the bridge itself, which could be seen very clearly from the protected boulder location. Sid had placed 12 men, or a reinforced squad, at that location and we were confident they could defend themselves and the bridge. They had their own radio for emergency communication with Sid and me, if needed, and a M-60 machinegun, in addition to their normal M-16 rifles and hand grenades. The troops we left there seemed confident that Sid and I had a good defensive plan before I started back down the highway. I asked Sid to stay with his defensive group that was at a bridge about one-half mile back down the mountain. However, he was just around a curve that blocked his view of the larger bridge location I had just left. As usual, radio silence was mandated which meant that no one was to talk on the radio except in a case of an emergency.

A couple of hours later, Sid broke radio silence and told me he could see tracer rounds going out into the valley, and he was sure they were coming from the large bridge location. I had my Company Radio Operator break radio silence again and try to call the large bridge location, but there was no response. I called Sid back and asked him to take immediate action to try to get to the location but reminded him of the dangers of moving on Highway One, even in the daylight, and this was a pitch-black night. With unbelievable speed, especially traveling on foot in the darkness, Sid and his troops he had at his ambush position moved to the large bridge location. He immediately called me and said he needed a medevac helicopter.

The chopper pilot did what chopper pilots always do, and in the darkness, on the side of a mountain, in a very small area that would have been dangerous to make a landing in the daylight, he landed and took out eight killed and wounded Marines. Why the gooks did not try to "blow" the chopper as it sat on the narrow road on the side of the mountain in the menacing darkness is unknown. It could have been because of their immense success while attacking the defensive position, and they decided to withdraw under the cover of the encompassing darkness.

In addition to our troops the enemy forces killed and wounded at the bridge, they captured one of our Company radios, the M-60 machinegun, and six M-16 rifles. "Chi-com" hand grenades that did not go off when they were thrown littered the area. The Chi-com grenade was a round-shaped piece of metal full of an explosive, and the metal object was attached to a wood handle about one-half as long as an average hammer handle. It enabled the person using it to throw it easily. Fortunately for us, the grenades, which were historically unreliable, did not all detonate, or our casualty rate could have been the entire 12 Marines at the location.

It was hard to believe it happened, but when I was on R&R in Hawaii to see my wife months later, I went to the Tripler Army Hospital to visit any of my wounded men who might have been sent there, even though I had no idea if any of them had been. I found one Marine who was at the bridge that night. He had lost his hand trying to throw one of the many Chi-com grenades away from himself and the Marine who was next to him behind one of the boulders. He confided to me that shortly after I had left the location that evening, some of the troops started smoking "pot," including the Corporal who we left in charge. He said the Corporal had assigned him a two-hour radio watch. He also said the radio was located at the back of the position next to the cliff where all of the loose material should have made it very difficult and noisy for the gooks to get to our defensive position. He added that at the end of his two-hour radio watch, he notified the Corporal that he was going to "crap out," or get a nap. He went on to say that no one took the radio watch, and that nine of the 12 Marines were smoking "dope." He related that the next thing he remembered was that there was an explosion where the M-60 machinegun was located and that the Marine in charge of the weapon was either wounded or killed instantly. He stated that at about the same time, he looked toward the radio and saw one of the attackers putting the radio on his back. His assumption was that some of the attackers had come up the steep embankment to where the radio was located, but because of the drug activity, no one noticed any noise the attackers may have generated.

I had already recommended that Marine to receive a bronze star for bravery for his attempt to protect his friend by throwing the grenade away, based upon information given me by Sid. The entire, tragic event was an example of the tremendous dangers involved with the use of pot in South Viet Nam. Because of this single tragedy, our enemy could now monitor all radio traffic within at least a mile of wherever that captured radio might be located in the future. In hindsight, it occurred to me that considering the thick, overhanging jungles along the highway, the plans we made and discussed with the troops before I left that location were probably observed. In addition, as fluent as some of the Viet Cong were in English, they could have even overheard our conversations.

A few days later, some of the village elders came to the perimeter to talk to "Honcho," meaning me. They informed me that "Linda had baby Marine," and I assumed they wanted me to notify the father. I told them there was nothing I could do. I did not know who "Linda" was, but according to the Gunny, there had been a Marine who apparently had some personal involvement with a local villager. However, he had rotated back to the world several months before. It was yet another reminder of the tragedy of Amerasian children. Amerasian children, in general, were rumored to have been the first to be executed by the North Vietnamese after they overran South Viet Nam.

The "Grapevine"

BEFORE I KNEW I WAS GOING TO COMMAND a Company without any experience as a Platoon Commander or Executive Officer, I had given considerable thought about the necessity of good communication within a combat unit. While I had never considered communication that in the criminal world would be branded as that of a "snitch," it became very apparent to me in the early days of my new responsibility that reliable internal information transmitted confidentially could be the difference between life and death.

One of my first priorities as the new Commander of Echo Company became the development of a "grapevine." I envisioned it as an informal way of receiving information of importance about our Company that would not only make the operation of the unit more successful, but also make it more secure. The Gunny had aggressively pointed out to me during my first night of command that it was crucial for me to know what was going on everywhere in the Company. He also impressed upon me that it was absolutely essential that everyone in the Company knew they could get information to me in a confidential way, and that I would take it seriously. He also advised me that any action I took needed to be commensurate with the seriousness of a situation, and if possible, I should always take enough deliberative time to make sure the information was correct before acting. Lastly, he said to make sure there was definitive action when I decided to remedy any proven allegation. He emphasized several times that any kind of informal, confidential information would only be forthcoming if those under my command believed I could be trusted, and that I would have to earn their trust. I told the Gunny that, realistically, I would expect this kind of information to come to me from the Staff NCOs and NCOs rather than individual troops. However, contacting me was to be an option open to every Marine under my command. He agreed, and I asked my Staff NCOs and NCOs to get the word out to my men.

The first information produced by my efforts came with the arrival of a new Navy Corpsmen. Corpsmen were in very short supply, so it could be jeopardizing the health and safety of the entire Company if we did not have enough of them. When this new medical person arrived, I welcomed him with "open arms." However, the first thing that crossed my mind was why he suddenly became available and was being assigned to my Company. 2/7 Command had recently told me that there were no new Corpsmen available, and that it was not known when any would be. I talked to him for some time, and we discussed not only his medical experience, but we discussed what units he had been assigned to since he arrived in-country.

One of the main points I expressed to him was that I had a zero tolerance when it came to the use of any drugs or alcohol. He said he agreed with me and assured me that he would never be involved with either. He had arrived in midafternoon, and before dark, I had information brought to me that he was carrying pot. I was able to verify the allegation, and I requested to have a chopper pick him up the next morning. My message to 2/7 was that I did not have the confidence in him to have him in my Company. 2/7 was not too happy about my decision, but about two and one-half weeks later, he was with an ambush as a member of another Company, and the ambush was discovered and attacked. He was killed along with two or three other Marines. The information I received from 2/7 was that drugs were involved.

The next instance happened on a Company Operation. We had been assigned a Forward Observer, or FO, for a 175 mm battery because we were going into an area that was so far inland toward Laos that no other artillery had the range to reach us. The Marine Corps did not have 175 mm artillery pieces. The weapon belonged to the United States Army and the FO assigned to my Company was Army personnel. Someone at 2/7 believed that it would be better if one of the members of that battery team was calling the fire missions when necessary. Keep in mind that this weapon had a maximum range of 22 miles, even though some of its accuracy was lost after a range of about 20 miles.

The first night after we had launched the Operation, Jack came over to me and said he was sure he smelled the odor of marijuana when he walked by the new FO's hole. We went over to where he was dug in, and while I visited with him, it was easy to conclude that his slurred speech and answers to my questions indicated that he was under the influence of something. I told Jack to take his radio, and I sent him back to 2/7 the next day. I notified 2/7 that I had relieved him from his duties, and the reasons why. Fortunately, we did not need the 175 mm artillery the night he was with us, but I was cautiously optimistic that we would have been able to call in the fire missions without the "whacked out" FO. Jack's experience as a police officer in Connecticut prior to reentering the Marine Corps to serve in Viet Nam was just one of the very valuable perspectives he brought to every situation.

The fastest information I received from the grapevine was when one of my Marines returned from R&R. We welcomed him back and listened to his stories about the great time he had. Within ten minutes after listening to his account of his time in the world, I had information that he had alcohol in his pack. I went to his hole and asked him if he thought we were going to "run short of water." He looked at me with a blank expression on his face and said, "No." I then asked him if he had anything in his pack that he thought I should see. He looked at me sheepishly and said, "Yeah. I guess so." He pulled out a good-sized bottle that was clear glass, and the glass was very thin. He handed it to me, and with just a casual glance, I could see particles of something floating in the liquid. I asked him if he wanted to get rid of it or if he wanted me to. He said he would, so I handed it back to him. He looked at it for a second and then, reluctantly, he smashed "the bottle of treasure" on the end of his hole. As I walked away, I hoped that I had prevented a possible safety problem for the Company, but by looking at the contents of the bottle, I may have saved his life. It was very nasty looking "stuff," including particles of something floating in the booze.

Because my men returning from R&R often had hard liquor with them, I decided to make a policy to help deal with the problem. After several discussions about the various issues involved with Jack and Sid, I became convinced that it was the pragmatic thing to do. The basis of my decision was that if they came back to our Company with any kind of booze, all they had to do was bring it to me and I would keep it in my bunker at Hill 190 on a ledge with their name on it. If they were not going out on patrol/ambush on a given night, they could come to me and ask me for a "snort" of their "private stash." I did not drink alcohol at that time in my life, and I did not think it was a good thing to do in a combat zone. However, I will have to admit, it seemed to help the situation considerably.

Aid and Comfort to the Enemy

TREASON IS THE ONLY CRIME enumerated in the United States Constitution. I had the opportunity to observe that crime being committed firsthand, in my opinion. From our position at the ESSO Plant along Highway One, we had a beautiful view of the DaNang Harbor. A small ship sailed into the Harbor one day. It appeared to be large enough to be "seaworthy," meaning it may have been capable of traveling in an extended ocean voyage. In this case, unbeknownst to us, this ship had traveled from the world to Southeast Asia, i.e., North Viet Nam. It did not seem to be anything that would be of interest to my Company, but we did notice that after it had been there several days, a South Vietnamese naval vessel took it under tow and began pulling it toward the ocean.

Just before darkness fell that evening, I received a message from Col. Love saying that someone on that ship being towed had jumped overboard and appeared to be swimming to shore. He also said the swimmer was thought to be a male American citizen, and that if he reached shore, he may try to climb up to Highway One. Col. Love made it absolutely clear that if that person was successful in reaching Highway One, I was to make certain that he was not wounded or killed. As I mentioned before, we deployed ambushes along Highway One every night while carrying out our mission of defending the many bridges on that transportation route.

I was not sure how I was going to carry out that order, but I could tell from the tone of the message, I would be best advised not to request clarification. After passing the word to my men, we were all amazed that we were being placed in a position to ask a potential enemy for their identification prior to opening fire. Unfortunately, this was an example of the politically driven logic required of military units in the Viet Nam War. As luck would have it, the swimmer reached shore and climbed up the mountain. He walked down Highway One into the position I was in. Fortunately, everyone in my unit responded just the way I was sure they would, and this "gentleman" was successfully apprehended. He identified himself to me as being Professor Butterworth who taught at Ohio University. He was a peace activist who was a passenger on the ship that we observed being towed earlier in the day. He and his group had been to North Viet Nam earlier in the month to deliver "humanitarian supplies" to the North Vietnamese to be used in their war effort against American and Allied Forces. His anti-war sympathizer group had come to DaNang to hold a press conference. Their request was denied as was their access to the docks in the DaNang Harbor, and they were told to leave. When they refused to leave, the South Vietnamese Navy began towing them out to sea. The Professor decided he would jump overboard, swim ashore, and walk to DaNang.

In his defense, I believe he was totally unaware of what combat troops even were. He was totally oblivious to the fact that military forces like ours were trying to protect the South Vietnamese people and provide support for the internal operations of their civilian society. His ignorance of the realities of life was stunning to me, and he was shocked that, as a college graduate, I was there serving my Country and "killing babies." He seemed totally surprised that we had the capability to have killed him instantly and somehow had avoided doing so. It was obvious to me that he thought war was a political process, and he had no understanding that real people on BOTH sides of the conflict were wounded and killed every day, INCLUDING Americans.

Several of my Marines requested about five minutes alone with him to give him an education that they thought he had missed out on in his protected life, but I had to decline, even though my men seemed to have an excellent point. Col. Love seemed very grateful when I notified him that we "had his boy." Our "guest" was picked up shortly after first light the next morning by a United States military vehicle from DaNang.

Despite our efforts to save his life, we did not receive a thank you note, and I apparently was not placed on his Christmas card list as a result of making his acquaintance.

Artillery and Fixed-Wing Support

PART OF THE MISSION REGARDING HIGHWAY ONE was protecting the civilian traffic on the dangerous trip up and over Hai Van Pass. On one particular day, the drivers of a small civilian convoy of trucks with bulk fuel tanks mounted on them entered the ESSO plant and loaded the tanks with diesel fuel. Back in the world, the vehicles would have been referred to as two and one-half ton trucks. Using the extremely dangerous Highway One, the convoy was going to haul the fuel over Hai Van Pass and down the other side to the towns of Phu Bai and Hue that were closer to the Demilitarized Zone, or DMZ. It was interesting to me that they were the same size trucks that we used on the farms and ranches in Wyoming. However, what was a Ford truck in the world bore the name plate indicating it was a Mercury, and the vehicles that were Dodge trucks in the world bore the name plate of a Plymouth. Highway One was the only road between DaNang and the Phu Bai/Hue area.

The convoy left the ESSO plant just before noon. We were not asked to specifically provide them with armed guards, but most of my Company was near the Highway at different locations at all times. The trucks had only gone about two miles when they were ambushed from enemy positions above the road. I think the attackers thought they were hauling gasoline and that the fuel would explode when they fired rounds into the fuel tanks. Because it was diesel, and because the attackers must not have used any tracer rounds, the fuel did not explode. The bullets went completely through the tanks, and fuel was squirting out both sides of the tanks onto the old asphalt road. All the civilians were able to escape, but it was obvious to me that the attackers were waiting for us to come to the assistance of the drivers of the damaged trucks.

I stopped the reactionary force I had with me a distance back from the ambush site so we would not be exposed to enemy fire from the rocks above the trucks. There was a significant danger if the fuel continued to squirt out of the tanks in the several small streams. If the streams of fuel had been allowed to continue to run back down the road and the fuel had become ignited, the road would have been seriously damaged. The ultimate result would have been that the road would have become impassable for a considerable period of time, and repairs would have to be made to the road by the Seabee Naval Construction Units once again. Trying to protect the Seabees while they worked in those very dangerous conditions was a huge problem for my Company.

From my agricultural background, I knew what the "butterfly" valves and "screw" valves on the fuel tanks looked like, and how they could be opened and closed. I decided to try to get to the four truck-mounted tanks that were leaking fuel onto the road to open the main valves so the fuel would drain much faster. If my "Country Boy" logic was correct, the increased volume would leave the actual road surface more quickly and drain over the cliff into the huge canyon beside the road. It also would eliminate the potential fire and explosive power of the fuel inside of the tanks on the trucks if it became ignited.

Crawling along the road was not a problem because I was down over the edge of the road, and it protected me from the small arms fire coming from the rocks above me. I was able to reach the rear of each truck relatively unnoticed and open the main product valves which sent approximately 2,500 gallons of diesel fuel from each of the four trucks cascading down over the cliff into the canyon below.

The next thing I needed to do was to get some artillery fire called onto the enemy positions above the trucks. That was the first time I had the opportunity to call for artillery fire in the mountains since taking command of my Company. I knew from my training at 2/7 Headquarters that the inaccuracy of the surface maps was a serious issue. With that in mind, I called for a fire mission that, on my map, should have made the first marking round hit approximately 400 yards east of our location in the canyon beside the road where the fuel trucks were sitting. The marking rounds contained white phosphorous, and were called "willy peter." When they impacted, they created a large cloud of white smoke so the impact location could be easily seen. In addition, the burning phosphorous also caused collateral damage to both humans and the impact area. If my plan was correct, I intended to make estimated adjustments of the following high explosive, or "HE", artillery rounds from the impact location of the willy peter rounds. The adjustments would hopefully cause them to impact on the target above the road. The final adjustment would have to be correct because, since my troops and I were close to the location of the actual target, there was not much room for error.

As we waited down the road from the fuel trucks and out of sight of the attackers, it took about 90-120 seconds for the first artillery round to reach us from the artillery battery location approximately seven miles away. We could hear the willy peter round coming for about five to ten seconds before impact, and the sound of the projectile did not make it appear to be headed for the canyon below. About three to four seconds before impact, I shouted for everyone to "Hit the dirt!" As we jumped off of the road, the willy peter round landed right where we had been standing... just a few feet from some of the leaked fuel. Fortunately, none of my men received any burns from the willy peter and it did not ignite the fuel. I imagine Charlie got quite a laugh out of that if he were watching, and the gooks might have wondered if I had suicidal tendencies!

I adjusted the HE round 150 yards to the left, and it came in much closer to the enemy positions. Then, with one more small adjustment on the next round, I called "Fire for Effect!" Because of the number of artillery pieces in the artillery battery that were available for our particular fire mission, we received 10-12 rounds of HE over the course of about 30 seconds. Hopefully, that was not a laughing matter for the gooks, and it gave them something to think about.

Right after the fire mission was complete, my aircraft Radio Operator told me that a flight of Air Force fighter/bombers were coming back from a mission north of our location and they wanted to know if they could help us with the ordinance they still had left on their planes. I told my Radio Operator I would be glad to talk to them.

There are two military principles that instantly became involved. The first principle was that the flight leader of the two aircraft was the Pilot In Command, or "PIC" of those two aircraft. Similarly, I was the unit commander on the ground, and I had the authority to accept or reject any support from any entity, including aircraft, artillery, naval gunfire, or another ground unit. I described the location of the target and we had a brief discussion of possible ground fire the flight leader could anticipate. I asked him to make his run on target from north to south, or vice-a-versa. The flight path of the planes and the trajectory of the bombs would have been across the target above us. He declined and said he wanted to make his "bombing run" down the slope of the mountain coming toward us. Since we were below the target, I was very concerned that some of the bombs might ricochet downward on the rocks, as they would sometimes do, and possibly land on top of my troops.

Even though the conversation was relatively short, it became heated. He made it clear that he could not land at his base without dropping his ordinance, and that he would have to drop it in the South China Sea if I did not allow him to utilize it on my target. I declined ungraciously, and the flight turned away to the east toward the ocean where they would dispose of their ordinance. I found out later that fighter/bombers in all branches of the services had to "pickle" unused bombs and rockets offshore prior to landing. What the exchange pointed out was how important Marine Corps Basic School Training was so that Marine pilots understood what it was actually like to be on the ground, and the necessity of a ground-unit commander to be able to decide how, or if, air support were to be used. Those U.S. Air Force pilots apparently had not had any similar training.

Promotion

By now, I still had less than five months in grade as a First Lieutenant. The Marine Corps was expanding in numbers and the 5th Marine Division was being formed at Camp Pendleton, California. Because of the need for more officers to fill the positions in the new Division...and because of the high casualty rates among Marine infantry officers in Viet Nam...and because I was filling a Captain's responsibilities, or " billet", I was about to be promoted to the rank of Captain. To be promoted so quickly was unusual, but my entire service in the Marine Corps to this point was far from what could be thought of as "normal."

Another Officer, who served in the Battalion Headquarters and was a Captain, was being promoted to Major at about the same time. A small promotion ceremony was going to be held at the Battalion Headquarters for the two of us. However, I was unaware of my pending promotion. I was simply notified one morning that a chopper would pick me up in the late afternoon and that I would not be returning that night. Since the message did not say why I was being picked up, I wondered if I had done something wrong...or worse. It turned out that I was flown to 2/7 Headquarters for the promotion ceremony, and the little ceremony took place in a "hooch" called the "Club."

The Battalion Air Officer had gone to DaNang and transported a "round eye," or American woman, who was a Red Cross volunteer, to the ceremony to "make sure things remained civil." Alcohol was available in the Club, and apparently much of it was usually consumed in those situations. Shortly after the ceremony, a rocket attack was launched against the 2/7 Headquarters. There was a bunker near the Club and we "dove" into it. Sitting there in the dark, I finally told the Red Cross worker that I was not "coming on to her," but I was sure we had met before. She said she thought so, too. We compared notes for a few minutes and eventually discovered it was her apartment in Washington, D.C. where my Officer Candidate friend from Texas became ill, and I had helped her clean up his "puke" and get the other Candidates out of her apartment. That was the first of several occasions where I would discover a prior connection with some of the people I came in contact with while in South Viet Nam.

After the rocket attack ended, a Marine asked if I would like to make a phone call back to the world to celebrate my promotion. I was astounded and said, "Sure," thinking it was probably a joke. It was for real, and he took me to a hooch where many of the larger radios were located. He explained that it would be the "wee hours" of the morning before we would know if it would be possible to complete a call. He explained further that it depended on how much the ionosphere cooled during the night. The radio signal had to bounce off the ionosphere for the transmission signal to reach the west coast of the United States. If it did, and if it were received by a ham radio operator, the call would be able to be transferred to a landline for transmission by telephone to someone in the world. I anxiously waited to give it a try.

On our first attempt, we were successful in completing the radio transmission to a radio operator in California. Because of the time differential, I assumed my mother might be home during what was the afternoon in Wyoming. I gave the radio operator my parent's phone number, and my mother answered on about the second ring. The California radio operator quickly instructed my mother that after either of us said something and we wanted the other party to respond, we should say "Over." I quickly said, "Hi, Mom, Over." She was so shocked, she said "Over." After she did not say anything else, I said "How are you? Over." And again, she said "Over." I said, "I love you, Over," and again she said "Over." Since the call could not exceed two minutes, it only consisted of short sentences from me, and her one-word emotional utterances of

"Over!" It was the thought that counted, and I appreciated the ham radio operator in California that gave me the opportunity to call home. I had a good laugh after I hung up.

My mother always loved to visit with people, but on that day, her vocabulary was very limited! After sleeping on the floor in the Club, I was flown back to Echo Company the next morning.

Elections

With Vietnamese local parliamentary elections scheduled in the fall of 1967, it remotely appeared that they were scheduled to coincide somewhat with similar elections in the United States. Through the efforts of the United States and South Vietnamese governments, all the villages and hamlets in South Vietnam were supposed to elect local leaders. The VC (Viet Cong) and NVA (North Vietnamese Army) threatened to kill anyone who agreed to run for elected office in those elections. Part of my Company's mission was to protect the Vietnamese people in the villages and hamlets in my Tactical Area of Responsibility, or TAOR, from the vicious VC and NVA attacks.

My TAOR was sizable and stretched for several miles to the south of our actual Company Base location. Because of the sheer size of my TAOR, protecting the people in it proved to be almost impossible from my Company's location in the Hai Van Pass region. Our daily efforts to protect Highway One and the many bridges over which it traversed consumed most of our time.

In the weeks leading up to Election Day, the enemy forces were successful in assassinating 12 of the 13 people running for the office of leading their hamlets and villages. In the world, we would have called the positions being elected as that of the Mayor. In my attempt to keep the last of the candidates alive until Election Day, I placed him in a boat and had it anchored offshore in the DaNang Harbor overnight. When the candidate came ashore on the morning of the elections, he was met by South Vietnamese troops who were supposed to protect him. As he walked up the beach, members of his "Security Team" wearing South Vietnamese uniforms fired 30-35 bullets into him at point blank range, and his bullet-riddled body was left there on the beach as a warning to everyone in the villages and hamlets that they were not to vote that day. As could be imagined, with all the candidates having been assassinated, the elections in our TAOR did not take place. In retrospect, it was probably a clandestine message to all U.S. military forces and our government that much bigger things were being planned by our enemies in the next few months, i.e., the 1968 Tet Offensive.

Hill 190

IN THE ENSUING WEEKS following the 1967 election debacle, my Company began a gradual, preliminary transition to a base camp called Hill 190. If and when it was completed, it would make our protection of the villages and hamlets west and north of DaNang tactically simpler by shrinking the size of the enormous TAOR we were currently responsible for. It was a cautious transition because, unbeknownst to me, military intelligence sources were indicating alarming increases in the number of enemy forces throughout I Corps. Our movements and base camp deployments were being calculated to correspond to the emerging NVA and VC threats. We continued to spend time operating north of Hai Van Pass as far as Phu Bai until December 28, 1967 when we were officially relocated to Hill 190.

The name, Hill 190, was a reference to the elevation above sea level referred to in meters. The Base Camp was located west of DaNang overlooking the Song Cu De River at the entrance to Elephant Valley. The large valley led from the coast of South Viet Nam westward into Laos. Elephant Valley was an important ingress and egress route for the North Vietnamese from the Ho Chi Minh Trail into the DaNang area. The Marine Corps military complex areas located at DaNang and Chu Lai were in the I Corps portion of South Viet Nam.

The mission for my Company when we arrived at Hill 190 had several components. The first was to prevent the VC and NVA from launching rocket attacks on the DaNang Air Base; the next was to continue to provide protection for the 13 villages and hamlets in our TAOR, while winning the hearts and minds of the inhabitants of these Catholic and Buddhist enclaves; and last but not least, we were to patrol known enemy routes of ingress and egress to the greater DaNang area of operations. However, my Company would regularly be called on to move to other areas of I Corps to conduct search and destroy operations.

Several issues pertinent to the warfare being conducted by United States Military Forces in Iraq, Afghanistan, and other areas in the Middle East had "very small roots" during our war efforts in Viet Nam. The I.E.D.s of today were what we called booby traps, which included manually detonated mines, the exploding of dud ammunition and bombs by the use of tripwires and crude "punji traps" (devices designed to mutilate, injure or kill anyone who comes in contact with them. They can be hung from trees as swings or placed in camouflaged holes in the ground and were usually handmade and very effective.) They were a constant source of danger which necessitated a change in the strategy we employed in our TAOR around Hill 190.

When one of the villages or hamlets was attacked at night by enemy forces, we utilized a reactionary force to come to their defense, and it had to be done very quickly. However, it was always dangerous because there were only a few alternatives available to us when we had to try to defend a particular location. I thought it was important to change our plan of reaction continuously so the enemy forces would not know which particular maneuver we might use.

On one particular night, we saw shooting start at one of the Catholic villages who had their own individual weapons, so we employed the use of a Six-by truck for the utilization of speed and surprise, while some of my Marines traveled on foot. In addition, we sent the truck through the rice paddies, which were relatively dry at the time, instead of going on the crude road which could have been anticipated. After the attack had been repulsed, our troops returned using alternative routes.

Early the next morning, we saw a jeep approaching our position. It turned out to be one of my Marines being returned back to us from his R&R trip. He was being given a ride by a Chaplain who was coming to our Company base camp for a small memorial service honoring some of my men who had been killed in action. As they bounced down the road between our position and the village we had helped defend a few hours before, there was an enormous explosion. The jeep was immediately blown into numerous pieces with the rear axle housing landing about 100 feet from the site of the explosion, and the engine block about that far in the other direction. The hole in the road created by the explosive device was about three feet deep and approximately six to eight feet across. Initially, we were not able to identify the remains of any of the individuals in the vehicle. We finally discovered a piece of our Echo Company Marine's body consisting of his right shoulder, neck, and head, and we were able to confirm that he had been in the vehicle. I could only thank God again that we had not sent the Six-by truck down that road the night before carrying some of our reactionary force.

One memorable thing about Hill 190 was a large rat that was my "roommate" in my bunker. He was the size of a small dog...so large that I called him "Mr."! One night I had dozed off leaning against the dirt wall. We had received a care package from some kind people back in the world, and, while most of the food sent to us was not edible by the time it got to us, I had been eating a small package of "raggy" corn chips that was sitting in my lap. When I awoke and felt into my lap for another bite of my chips, the bag was still there, but the chips were gone. In the filtered, dim light of my flashlight, I could see that my "friend" had eaten the chips out of the sack in my lap.

I decided that was enough, so a few days later I was able to get a large "varmint trap" with some re-supplies. I baited it one night with a piece of caramel, which survived some of the heat because it was individually wrapped, and anxiously waited for my "roommate" to return. It wasn't too long before I heard the trap snap shut. My dimmed flashlight revealed the trap had caught the culprit with the caramel still in his mouth, but he was looking at me with a look of defiance. In a few short seconds, he "shrugged" his shoulders and pulled his head out from under the heavy, spring-loaded trap. He walked away chomping on the caramel. From that day forward, I always called him "Sir."

As I mentioned earlier, pot smoking was an ongoing problem. Because of my conservative upbringing in rural Wyoming, I could not understand the use of illegal drugs. However, in my Company, the problem was real and it had to be addressed daily. One incident was reported to me by Jack. Early one morning, he came to me and said he wanted to show me something in the 81 mm mortar pit. The pit was an area surrounded by two or three rows of sandbags that contained the mortar tubes, mortar rounds, and the powder increments that explode in the mortar tubes causing the round to be launched from the tube. There were several holes burned in the sandbags indicating someone had been smoking in an area where it was clearly a "near suicidal thing" to do. Other evidence indicated it was not the average cigarettes that came with our supplies. It was obvious we had a problem that we had thought was under control. It was also an indication that I had to do more to restrict the ability of my men to have access to the pot that was so available in our area of operations.

Racial problems were enormous in the world, and they were real in Viet Nam. While I believed we had good race relations in my Company, a very militant Marine arrived one day to become the newest member of Echo Company. I assigned him to a platoon, and the next day, he was involved in a fist fight with his squad leader, who was a Corporal. The Corporal had been in-country for several months and was one of my best NCOs. The Corporal was smaller than the new Marine, and the Corporal was white, while the new Marine was black. According to the Corporal and his squad members, the squad was filling sandbags and the new Marine refused to work. The Corporal gave him an order to help with the work, but he ignored the lawful order and then he punched the Corporal in the face. A fight ensued and ended with the Corporal pushing and shoving the Marine over to where I was. There was not a "winner," and in reality we all lost. When I tried to talk to the Marine, he became very belligerent and insulted me personally.

I immediately asked Cathi to request a helicopter, and when it arrived, I sent the new Marine back to Battalion Headquarters with a message for Col. Love to "Run him up and disc him." That meant I wanted the

Marine punished to the fullest extent possible. I was told that upon his arrival at the 2/7 Headquarters, he appeared in front of Col. Love. The Marine hurled an insult at our Battalion Commander, which was the wrong thing to do. He was immediately sent to the Marine Brig, or prison, located close to DaNang.

Any time spent in the Brig while in South Viet Nam was considered "bad time," and it did not count as part of the 13 months Marines had to spend in-country. In addition, the Brig was at the same location as the wire-fenced Prisoner of War facility where most of the VC and NVA who had been captured were held. I was later informed that when the Marine arrived for his prison stay, he was told to stand at attention. He refused to do so, and he insulted the Marine Guard who immediately "butt-stroked" him in the mouth with a rifle. As the story went, he laid on the ground for a minute or two, and after spitting out some broken teeth, he stood up...and stood at attention. When he was returned to my Company several months later, he had a considerably different attitude than when he joined Echo Company. However, as I will relate later, what happened the day after his return was unbelievable.

Col. Love sent me a message that said we were going to be deployed on a search and destroy operation, and we would be relieved at Hill 190 by another Company while we were gone. The message went on to say my Company would be moved back to the Battalion Headquarters for the briefing and deployment. After we arrived, we were notified that the operation would be postponed for three days. It provided an excellent opportunity for my troops to get some badly needed rest. An F-4 Phantom jet pilot I met while I was in Quantico was stationed at the Chu Lai Air Base south of DaNang. I asked Col. Love if I could try to get down there to see him for a day. Col. Love agreed.

A Dream Came True

I HITCHED A RIDE TO THE DaNANG AIR BASE and jumped in the back of a C-130 that was bound for Chu Lai, along with about 100 Vietnamese civilians. We sat on the floor with the back ramp of the cargo bay open for the 15-minute ride. Upon arrival in Chu Lai, I went to the Air Wing Headquarters and inquired about Major Jackson. They said he was airborne and would be back later in the day. I found some shade and "crapped out" until he arrived just before dark. We had a good visit, and he offered to make good on his promise he made to me at Quantico to take me on a mission in an F-4 if I ever found him when we got to South Viet Nam.

Prior to getting some sleep, the Major showed me a report he was having to file because his plane had been hit by ground fire a few days earlier while he was flying air support for the Marines at Khe Sanh. The Marines there had been under siege for several weeks. The North Vietnamese were convinced that if they could defeat us there, it would have had the same effect on us as it did on the French when they were defeated by the Vietnamese at the battle of Dien Bien Phu. The French loss led to their withdrawal from Indochina forever.

Major Jackson's plane had been hit by ground fire causing the fuel tank under one wing to explode resulting in serious damage to his F-4 Phantom aircraft. Somehow, he managed to fly the damaged plane back to Chu Lai, and in the darkness, he was able to hook the morest landing cable on the end of the runway that was used for emergency landings. He said it is similar to the same cable system used for landings aboard aircraft carriers. The damage to the electronic systems on his plane had prevented him from being able to "pickle" his ordinance in the South

China Sea, that I mentioned earlier was the standard practice. After all of the display of personal courage, the skilled, professional flight and control of a damaged aircraft being returned to base, and after he made an unbelievable landing without damage to anything or injury to anyone, he still had to complete numerous pages of a report to explain why his plane had been damaged.

He said the accepted operational procedure for aircraft flying close-air support for ground units was to make one pass over a target, dropping the entire ordinance at one time. He also said that any time a plane was damaged or shot down making more than one run on a target, it was mandatory that the pilot file the "bureaucratic report." From a ground-unit commander's perspective, we always preferred the pilots to drop one bomb, or "singles", at a time. That allowed us to have the damage to the target spread over a larger area and we could give the pilots information so they could possibly adjust their next run, depending on what we were seeing on the ground. In every instance that I saw as a ground-unit commander, pilots were always willing to defy their regulatory protocol if I asked them to, even though it might jeopardize their personal safety and risk their being subject to possible disciplinary action. I was immensely humbled by that revelation.

It is very understandable that a plane that makes three or four passes over a target is more apt to be hit by enemy fire than it is if it makes one pass. He explained that pilots are "creatures of habit" in certain ways, and that enemy forces would often watch a pilot make his first run on a target noting his angle of attack. They would adjust the sight on their .50 caliber machineguns, in particular, to reflect that pilot's angle of attack on his initial approach to the target. When that pilot attacked the same target again, enemy gunners could use the pre-set angle on the sight to make their firing at the plane more accurate when leading the target to compensate for the speed of the aircraft. Keep in mind that these outstanding pilots were working amidst target markings, enemy fire, and conversations with ground-unit commanders and Forward Air Controllers...at speeds of 300-500 knots per hour and traveling as much as 5-9 miles per minute, which is one mile per 6-12 seconds.

The bombs they were dropping for us were often "snake eyes." Those bombs had four flat metal arms on them that would swing out after the bomb was released from the plane. The arms would create enough drag, or resistance, against the air so that the bomb would slow down. This allowed the plane to get far enough away from the impact area before the bomb hit the ground so the plane would not be damaged as a result of the impending explosion. It never ceased to amaze me that, while the planes were traveling at those tremendous speeds, they were dropping the bombs when the planes were approximately 50-100 feet above the ground. The pinpoint accuracy of those aviation heroes will live in the annals of military history forever, and ground-unit commanders like me will be forever grateful to them.

The next morning, we were at the end of the runway going through the pre-flight checklist when it finally "rung in" to me that I was realizing my life-long dream of flying in an F-4. I was not flight qualified, I had never been in an ejection seat, I had never worn an oxygen mask, and I had never worn a "G-suit." The chances that the Air Wing and Squadron took for me to take that ride will never be forgotten. I was sitting in the backseat where the Radar Intercept Officer, or R.I.O., normally sat. We were to be the Flight Leader of a two-plane sortie on a mission to make a TPQ-10 drop over North Viet Nam. A TPQ-10 drop was a high altitude, radar-controlled mission where all bombs are released at once followed by an immediate right "break," or turn, toward the South China Sea. Everything went as planned and we turned south toward Chu Lai once we were safely past the coastline and "feet wet." Major Jackson took us quickly over the speed of sound so I could experience that and add it to this unbelievable "dream-come-true."

With my lack of the proper qualifications, my flight was not to include any kind of close-air support, which is low-level flight in support of ground troops that I described earlier. However, the South Koreans, or ROKs, west of Chu Lai in the Go Noi Island area were in need of assistance on our return. The decision was instantly made to provide assistance to our Allies. Since all we had left were air-to-ground rockets, our first run on target started at about 8,000 feet at a speed of 500 knots. As we roared toward the target, I could feel my body forced against the seat restraints when the first set of rockets launched. As soon as the launch was completed, Major Jackson hit the afterburner and we went into a nearly vertical climb. I saw the G-meter in front of me on the instrument panel go past 6.5. That meant we were pulling over six and a half times the normal force of gravity as we went through the bottom of the dive and began to climb.

My vision blacked out and I "came to" as the Major rolled us over on our backs as he prepared for the next rocket run. The ground crew at Chu Lai had given me a small plastic bag to use in the event I got sick during the flight...and it was a good thing. As we rolled in on the target the second time, fired the next set of rockets, "hit burner" and went through the G-forces, I was "out like a light again." This time when I came to as we rolled onto our backs for yet a third and final rocket run, I quickly pulled my face mask off and made good use of the plastic bag. As we were turning in on the target, Major Jackson shouted at me over my headset to get my mask back on because it was making a loud noise in his headset. I made a muffled, "I can't; I'm sick," and figured it was better for him to turn off my microphone and headset than it was for me to fill my mask with what was supposed to go into the plastic bag.

The ground crew had seated me in the back of the F-4 before takeoff, and they also connected a rubber hose inside the cockpit to a piece of equipment that fit around my thighs and abdomen. They called it a G-suit, which, as I mentioned earlier, was just one of several things I was experiencing for the first time. It is designed to instantly create pressure on those two areas of your body to keep the blood from being forced downward away from your brain during periods of increased "G-forces." As I was watching the G-meter in the bottom of the dives, the suit was supposed to inflate when the meter started to rise to counteract the effects of the multiple G-forces. I remember excruciating pain in the bottom of the dives, and I felt like I was being forced through the bottom of the seat and the plane, but I just assumed I was not tough enough for what I was experiencing.

When we landed a few short minutes later, the ground crew helped me out of the plane, and as they did, one of them reached in to disconnect the G-suit. The crewman noticed that the hose was dangling loose and he asked me if I had disconnected it. I told him I had not. He said, "Oh my God, man. You were sitting in there without any G-suit." That may explain some of the problems I experienced during the "rocket runs."

In the preflight mini-briefing, they had also explained the ejection procedure, "Just in case you need to, but you probably won't." I definitely wanted to know how to do it, and they explained that there were two handle grips just above my flight helmet that I could pull downward which would pull a curtain in front of my face, resulting in an instantaneous ejection. The second way out was a small switch that was located on the seat between my legs. It could be used if the G-forces at the time the ejection was needed were too powerful, preventing me from being able to get my arms above my head. In the post-flight debriefing, amongst all the laughter about me "blowing lunch," they asked if I had remembered what they had told me about a possible ejection. I told them I did and that I kept one hand very near, or on, the ejection button between my legs. They gasped in disbelief explaining that the switch on the seat between my legs was extremely sensitive and that I was very fortunate not to have activated it, especially when I was blacking out.

It was only then that they told me the week before my flight, a pilot and his R.I.O. were returning from a night bombing mission over North Viet Nam. There was a mechanical malfunction causing the canopy over the pilot's seat to come off. The backseat in the Marine F4s normally did not have any flight controls. When the front canopy came off in the darkness, the R.I.O. immediately "punched out," assuming that something had gone wrong and the pilot had ejected. When the pilot landed in DaNang, he looked in the backseat and, of course, the R.I.O. was not there. The R.I.O. was rescued some hours later in the South

China Sea, "swimming with the sharks." Once again, I was reminded of the old saying that, "It is always better to be lucky than good, that way no one expects it."

Man's Best Friend

AFTER THAT "EXPERIENCE OF A LIFETIME," I caught a hop back to DaNang late that afternoon and arrived back at the 2/7 Headquarters just before dark. Col. Love briefed me about a search and destroy mission he was having my Company launch the next day. He told me we would have the use of a Scout Dog Team that would be attached to us for the duration of the mission. These Scout Dog Teams consisted of a trained dog and his Handler. In addition, it was necessary for me to assign one of my Marines to assist the Handler. These dogs were trained killers, and they were so aggressive, no one could approach either the dog or the dog's Handler without the Handler first giving the dog certain restraining commands. The dogs were used to warn us of ambushes, smell and alert us about munitions and "punji-traps," and they could capture an enemy combatant without killing them simply by being given a single command from the Handler.

One of the purposes of the Marine assisting the Handler was to have someone along to carry water and food for the dog. The rest of my Company would split up the gear belonging to the Scout Dog Team Assistant so he would not have to carry so much weight. The Scout Dog Assistant always carried five gallons of water for the dog, which is 40 pounds by itself. The other reason a Marine assisted the Handler was if the Handler were injured, wounded, or killed, someone would have a small amount of familiarity with the dog that could allow him to, hopefully, control it. This was necessary because if the dog ever turned on one of my men, we would have had to destroy the magnificent animal. An example of how aggressive these dogs were was when one bit and ripped open the thigh of my artillery officer when he ran by the Handler and his dog during a "firefight."

I meticulously briefed the Handler concerning my expectations of him, and he did the same with me. I told him that everyone in the Company, including him, was required to carry a two-pound block of C-4, a plastic explosive that had a malleable, "cake-like" composition. The C-4 was sealed inside a clear plastic container. Because of its malleability, a blasting cap inserted into it and a fuse connected with an igniter was all that was needed to activate it. It was a very unusual material in that if you pinched off a very small piece of it, you could set it on fire without it exploding. Because it burned so easily and the combustion generated so much heat instantly, Troops were always tempted to use it to heat their C-rations. I had strictly prohibited the use of it in that manner, and I included that prohibition in my briefing with the Handler. The most important reason was because we had numerous needs for the explosive, often in ways we may not have expected. In addition, the bright glow emitted when it was lit was visible from a considerable distance. He said he understood.

In the next hour or two, we were airlifted to the jumping-off point for that particular search and destroy mission southwest of DaNang. Since I had not been afforded the opportunity to do any reconnaissance, we were relying totally on my briefing from Col. Love. When we were dropped into the area to pursue possible enemy forces, our progress was slow and deliberate, and we found a good defensive position for the night and began digging in.

It was just about dark when the Dog Handler brought the Scout Dog over to my hole and asked me to look at him. He was a beautifully marked silver/grey and black German Shepherd, which most of the Scout Dogs were. I always wondered why they used dogs with the longer hair in the heat of the jungles, but there must have been a reason. Because of the heat, a Scout Dog could only work for about two hours at a time before needing relief, and the longer hair on the dog seemed to exacerbate the problem. Since the Vietnamese in the countryside ate their dogs, we used to joke about our Scout Dog getting captured and fed to me at the next celebration at one of the villages or hamlets. After the first celebration where I had to eat the raw fish called "gukma," rice that was very rank and rancid, raisin wine, and possibly some dog meat, I decided that the dysentery I experienced for the next two weeks was an experience I would try to avoid.

This particular Scout Dog was there to assist us and hopefully help us avoid getting some of my men wounded or killed. As the incredible animal stood there, I could see his muscles twitch ever so slightly every minute or so. I asked the Handler if he was alerting. He said no, and then he finally revealed what happened. He said he used some of the C-4 he was carrying to heat his C-rations... just exactly what I had told him not to do. He said he set the open container of C-4 down on the ground while he began digging his hole. The Handler went on to say that when he looked around a few minutes later, it appeared the dog had eaten approximately 25% of the explosive material, which would mean approximately one-half pound. I told him that I was sure there was lead in the explosive material and, if that were true, I was certain the dog would begin to suffer lead poisoning. The only thing we could think of was to get the dog to vomit, but we only had medications that would stop vomiting in humans. I finally told him to keep an eye on him and returned to digging my hole.

It was not long until I heard a commotion, and my men passed the word with loud whispers to have me go where the dog and Handler were. When I got there, the dog was lying on its side completely rigid. In a somewhat similar fashion as I did with newborn calves during calving season on our small ranch in Wyoming, I jumped down in front of the dog and cupped my hands around his mouth. I told the Handler to do chest compressions from the top and I began trying to blow air into the dog's mouth. After 20 or 30 seconds, the dog let out a big sigh and started breathing. Shortly after, he sat up panting.

The dog Handler was frantic and told me I had to call a medevac for the dog. He went on to say it cost $15,000 to train the dog, and he would be court martialed if the dog were not given the best care we could give it. I told him I would call for a medevac helicopter, but I was very doubtful that one would be sent. In addition, I told him to try to give the dog resuscitation if he had any more seizures. As I had suspected, the Marine Air Wing said that they were not going to risk a million-dollar aircraft and its crew to come out on a pitch-black night in "Charlie Country" to pick up a dog.

The dog had two more seizures after the initial one and it did not survive the last one. It was sad to see the Handler carry his dog, wrapped in a poncho, onto a helicopter the next morning. I filed a report with Battalion Headquarters, and we did not hear anything about disciplinary actions taken against the Handler, if any.

The Scout Dogs seemed to suffer some of the same mental trauma as combat Marines. When we were engaged where there were bullets going past us, most of the dogs would bite and snap at the air. It was apparently because the speed of the rounds made them break the sound barrier as they went past us resulting in a "popping" or "cracking" sound. The Handlers had to keep a strong hold on these aggressive warriors because, as I noted earlier, they would bite anyone who came close to them in the noise and confusion.

One particular Scout Dog had spent some time in Khe Sanh during the NVA siege of that Marine Base Camp in the fall of 1967. Whenever he heard gunfire or explosions, he would fall over, his eyes would roll back in his head, and he would urinate on himself. We sent him back to the rear area, and we were told later that the dog was apparently "shell shocked." The dog was sent to Okinawa for a complete rest. The prognosis was supposedly 50/50 that he might ever recover.

On another small Company search and destroy operation, we had a Scout Dog leading us on a jungle trail when he suddenly alerted. I was behind the platoon that was in the lead on the trail, so they called back and told me that the dog had alerted. I told the Platoon Commander, who had been with us for about two weeks, to wait until I got to where the dog was. The Platoon Commander and the Handler decided to move forward a small distance while they waited for me to get through the dense jungle trail. I was about 75 feet from them when there was an explosion. I took a few more steps forward and the dog came charging down the trail with his leash dragging behind him. His entire lower jaw had been blown off. The dog Handler was killed, and the Platoon Commander was wounded.

The dog had alerted them to what was a claymore mine. They were usually set up along a trail, and normally at ground level. The mine was slightly curved to give the explosive power a fan-shaped field of lethal damage and they were normally set off manually. Some of my Troops down the trail caught the dog and he was medevacked with the wounded Platoon Commander and the body of the dog Handler. Our Corpsmen bandaged the dog's head the best they could to control the bleeding, and we did not think he would survive. We found out several months later that there was an article in the *Stars and Stripes* military newspaper, reporting that the dog had made it to Okinawa and that the veterinarians there had successfully built the dog an artificial lower jaw.

These animals were some of our greatest heroes in the Viet Nam War. They were "drafted" to serve in a war of which they had no previous knowledge or experience. They trained to be the very best they could be, while being motivated by a desire to honor and serve their leader. They did not question a command and were so loyal and disciplined that they only took commands from their trusted best friend, the Handler. In addition, they performed their duties with courage and perfection, regardless of the danger of injury to themselves, even if it meant giving their lives for their Handler and their Country. I hope that as a result of this writing, readers will have a greater understanding of what the Scout Dogs did for those of us who served in the combat units, and the great debt our Country owes these amazing animals. There is no way of estimating the number of lives saved by these fearless warriors.

After having spent my entire life around all kinds of animals, I will be the first to admit that I do not fully comprehend the place of animals in the food chain for humans, nor do I claim to fully understand what animals think and do. However, Jane Bakalian, a wonderful friend of mine for many years, has taught me innumerable lessons about communicating with animals, especially horses. She is a model example of how communication can exist between humans and animals in a very special way, and the interaction with animals she demonstrates should be pursued by more humans. As with these Scout Dogs, I believe the results of the proper interaction with animals are some of the most rewarding experiences we can have in our lives.

Cool, Clear Water

IN THE ANNALS OF WESTERN MUSIC, popular in the western region of the United States where I was born and raised, there is a musical ballad written by Bob Nolan in 1936 that had water as the subject of the song. Though it may not sound very exciting, the words of the song came to my mind frequently when we did not have enough water, in South Viet Nam, which was not unusual. The words begin, "All day I've faced the barren waste without the taste of water, cool water. Old Dan and I with throats burned dry, and souls that cry for water, cool, clear water." Where we were in the paddies and jungles, there was no potable water. The water wells were all hand dug, so by their very nature they were very shallow, and the water in them was contaminated by any reasonable standard. The shockwaves in the ground generated by bombs, artillery, and rocket fire would either collapse the wells or shake so much mud, plant residue, and pollution loose from the walls of the wells that the water was unfit for human consumption.

However, it was all the water the indigenous population had, unless they drank out of the rivers. It was not like the Vietnamese could boil their water to purify it. The NVA and VC would throw dead bodies, both animals and human, in the rivers, streams, and sometimes even the wells to pollute them because they were aware that we had to use them for water supplies much more often than we would have liked. Looking back, it is amazing that we did not have more cases of typhoid fever and other waterborne diseases. Maybe all those vaccinations in Okinawa were worth it.

We tried to be resupplied with water as much as possible, but the reality of it was that we could only carry so much water with us, considering the 40-50 pounds of armament, ammunition and food that we already had to carry. The halazone tablets we carried to purify water were so distasteful that the river water looked better for drinking on many days. On one occasion, we attempted to have water brought to us in canisters used to ship powder increments to the artillery units. My reasoning was it would be delivered in waterproof containers, and if they were dropped into our positions as part of a resupply, the containers would not rupture when they struck the ground. However, they did not wash the powder residue from the inside of the containers, and that effort was not replicated again. It made me think daily how fortunate we were in our Country to have an abundant supply of cool, clear water.

We seldom had the chance to wash in any way, except when we were in base camp, and then it was on a limited basis. During the dry season, it was not unusual for us to go for long periods without the ability to wash. There were times when we could roll slime off our arms that was nothing more than accumulated body oil and filth and grime from the jungles and rice paddies. The stench of our bodies was much worse than we imagined, I am sure, but we all smelled the same, so it was no big deal. Often my map would indicate a stream in the mountains, and when we would get there, it would be dried up. However, on one occasion, we came across a mountain stream that had enough water in it that it formed several shallow pools as it flowed through the rocks. As you might imagine, we put out security and, to the last man, we took turns lying in the water and soaking our faces. It took some time out of our search and destroy mission on that one day, but it was one of the most refreshing things I can ever remember in my life.

Moving on with the Mission

THE MILITARY TACTICS OF THE NVA AND VC appeared to be classic guerilla warfare. They tried to avoid contact with us until they were ready and believed they had an advantage. They were skilled in strategically retreating when attacked, and then attacking us when we were moving in a different direction, maneuvering, or changing missions. When they did stand and fight, it was when they thought they had military superiority either in numbers or weaponry. Mother's Day, 1968, was one of the days they chose to initiate a large-scale attack on my Company during Operation Allen Brook which I will relate in a later chapter.

On a particular search and destroy mission in which we were engaged, enemy contact was sporadic. On the third day of the mission, we had an accidental discharge, and unfortunately it was not the first. A Marine who had just arrived from the world came in thinking he was "John Wayne," as some young Marines did. With 150-200 young "killers," all of whom had weapons, it was a constant safety training exercise to prevent accidents, and many of the accidents involved serious injuries or death. We had been dug in all night in our defensive position, and when we moved out the next morning, Marine "John Wayne's" platoon was going to be bringing up the rear, so he was still in his hole as we walked by. I normally had four Radio Operators within a few feet of me in different directions, even though they were usually spread out. As we were leaving the perimeter, we were walking single file with the Radio Operators immediately behind me.

The M-16 rifle, which was the primary weapon for combat units, had a three-position selector switch on it, near the grip; one position was for safe, one for automatic weapon discharge, and one for semiautomatic, single-round firing. It was a problem of immense concern because oftentimes Marines would pick up their rifles and pull the trigger to see if it was on safe, instead of looking to see. The problem was so acute that I had issued an order to check their weapons visually to see if it was on safe, with no exceptions except if we were under attack. The M-16 would fire 750 rounds per minute on automatic, which is just over 12 rounds per second. One careless pull of the trigger would usually empty the magazine, which was supposed to hold 20 rounds.

"John Wayne" picked up his rifle as we were walking by and pulled the trigger to see if it was on safe. A burst of 15-20 rounds were fired instantly. He was so close to my Radio Operator who was immediately behind me that very small metal fragments from the rounds coming out of the barrel of the rifle lodged in the Radio Operator's buttocks. Unbelievably, the rounds barely missed everyone else who was moving out. Immediately, I had to decide what a Rifle Company Commander should do in a war zone when one of his troops just endangered his entire unit, and caused minor injury to another Marine due to his careless, personal conduct. In addition, he had disobeyed a lawful order not to check his weapon that way.

We did not have time to have a full-scale discussion concerning the situation, nor could I afford the time to discuss it with Sid, Jack or others who might have a different idea. As in many combat situations, a decision had to be made quickly. Because it was such a serious issue, I decided to make an example of "John Wayne." I took his weapon and gave it to his Platoon Commander. I told him that he was to travel all day without a weapon...but privately I told the Platoon Commander to get it in his hands immediately if we had enemy contact.

Anyone who has been in a military combat unit can imagine the consequences of my actions if he had been injured or killed that day. However, our mission in the war zone was to kill as many enemies as possible. I believed that careless actions by my men that endangered the safety of the rest of the Company could not be tolerated, especially if it led to the wounding or death of our United States forces. Discipline often has to be administered swiftly to avoid jeopardizing the mission, and I believed that was one of those times. I will readily admit that I issued several prayers to God throughout the day for the Marine's safety, and my prayers were answered. He appeared to have learned a lesson from the incident, and I wanted to make sure every other Marine in the Company did too.

I would like to report that the M-16 safety issue never happened again. However, that same night, when we were digging in, the Corporal who was the Weapons Safety Trainer for my Company unbelievably did the exact, same thing. This time he had the rifle pointed in the air and it was on semiautomatic and he only fired one round. So... I had a quick conversation with myself and decided to order him to start digging...and not to stop until first light in the morning. I am sure the "bad guys" were very puzzled about what was going on and may have decided to leave us alone for the night, since someone digging a hole all night in the rocky soil of the mountains where we were must be some kind of trick. We did not have to worry about anyone in our Company sleeping that night with all the clinking and clanking of the Corporal's entrenching tool as it struck the rocks. As in the many decisions I had to make as a Company Commander, I made some that were wrong, and maybe these two were some of those. However, we did not have any more M-16 accidental discharges as long as I was the Commander of Echo Company.

Prior to this, I had other incidents when the Marine involved said it was an accidental discharge. One was while we were still providing protection on Highway One and guarding the ESSO plant. No one witnessed the "accident." The Marine involved said he was starting to sit down with his M-16 in his hands, and the first finger of his left hand was on the trigger and the first finger of his right hand was over the end of the barrel. He said the barrel was pointing straight up. He said as he sat down, the finger on his left hand accidently pulled the trigger and shot the end off of his first finger on his right hand. The rifle was on semiautomatic and only one round was fired. He was right-handed so the finger that had the end blown off was his "trigger finger." When the Corpsman brought him to me, I could tell the Corpsman was dubious as to what had happened.

The Marine said he needed to be evacuated to have his injured finger treated. The injured finger had been placed so exactly over the end of the barrel that when the rifle fired, it separated the end of his finger exactly at the first joint of the finger. It almost looked like it had been surgically removed. After conferring with the Corpsman about the possible medical issues involved, I told the Corpsman to bandage the finger. In addition, I told the Marine he was staying with the Company and that he would not be medevacked. When he complained that he could not fire his weapon without that finger, I told him to use one of the other three fingers he had left. We did not have any more of those particular issues either.

One of the most horrific cases of an accidental discharge that I was aware of happened at the Battalion Headquarters. The Corporal who was the .45 caliber pistol Weapons Safety Instructor for the Battalion was attempting to clear his pistol and made a tragic error. The proper way to clear the .45 caliber pistol is to bring it to the raised pistol position, which is to point the pistol upward at a 45-degree angle and away from everyone else. Then the magazine is to be removed, followed by pulling the slide back to eject any round that may be in the chamber. The person clearing the weapon is then supposed to physically look into the chamber to make sure there is no round in it. If the chamber is clear, the slide should be allowed to go forward. The final step is to pull the trigger to let the hammer go forward with the weapon still pointed upward at a 45-degree angle.

The Corporal made an unthinkable mistake and did what several hundred times he had taught others not to do. He grabbed the slide from the front of the pistol with it pointed into the palm of his hand. He pushed the slide back until the barrel hit his palm and, realizing he had erred, let the slide snap forward chambering a round. Since the hand he was using to hold the pistol was gripping the weapon and his trigger finger was griped against the trigger, the weapon discharged when the slide and trigger snapped forward. The round went through his hand and into the chest of his best friend who was standing beside him, killing him instantly.

Col. Love appointed me to sit on his court martial proceeding. We reduced him to the rank of Private and allowed him to serve the rest of his tour in South Viet Nam. He was so disgruntled, he later was seriously wounded when he appeared to be acting out a suicide attack on an enemy position. As a result of his wounds, he was medevacked back to the world. It was a tragic end to the lives of two excellent Marines...one dead and one who had lost the will to live because of his mistake.

As we proceeded with this particular mission following the two accidental discharges, the next day was uneventful, except for a message I received that another unit would be in our area of operation and it would be primarily an armored unit. I was not given any radio frequencies to use in case I needed to contact them, and the tone of the message was that they would be moving rather quickly. I have mentioned the cunning and creative tactics of the VC and the NVA, and my contention was confirmed that day.

We had been moving in single canopy jungle and brush for most of the morning without any contact, and we stopped to check conditions. In just a matter of minutes, we could hear what sounded like the armored unit going past us a distance away in what must have been an area of less foliage. In what was a classic act of deception, a small enemy unit was able to get between us and spray the armored vehicles, mostly tanks, with small arms fire. The armored vehicle personnel apparently did not know we were in the area and they began firing their machineguns into the foliage where the automatic weapons fire had come from. When we had stopped moving just minutes before, everyone in the Company had taken a prone position as we almost always did. When the armored unit began firing, they were riddling our position with hundreds of rounds. We could only hug the ground. When the firing stopped and the unit moved on past us, we quickly checked everyone in our Company because it seemed that there would be casualties, and possibly numerous ones. Unbelievably, no one was even wounded. However, when I reached for my canteen on my hip a few minutes later to get a drink of water, I found it was empty...and it had a bullet hole through it. Once again, we could only thank God for sparing us any injuries or deaths that day.

That night I received a message that we would be airlifted the next morning to pursue an enemy force closer to the area known as Happy Valley. We were dropped into a "hot LZ," or landing zone, which means the helicopters moving us were taking enemy fire while we were approaching the LZ and as we jumped off the choppers in the landing zone. There was high ground surrounding the LZ. Again, I had been unable to conduct any reconnaissance prior to our landing. Since no one had any idea of what we were getting into, I had gone on the first chopper with my Radio Operators and some troops. It was one of those "drops" where you could actually see a tracer round now and then going through the chopper as they came in one door and out the other. The rate of our descent seemed much too fast. For a few seconds, I wondered if the pilots had been hit, but when we narrowly missed a huge pile of boulders just before we hit the ground, I assumed the chopper was still under control. As we jumped out the back of the CH 46, dirt, rocks, grass and what later appeared to be small pieces of rotor blades were flying around like we were in a tornado.

We were told later that the chopper had apparently hit the ground so hard that the rotor blades had actually grazed the ground. If that was the case, I still cannot believe that the pilots were able to fly out of the landing zone. However, it was just one of many miracles that helicopter crews did on a routine basis. The problem for us was that we were still there. More choppers followed with more troops from my Company, and when the insertion, or "drop," was completed, we were able to get the upper hand and the enemy withdrew.

We climbed up on a ridge to dig in for the night and tried to report back to anyone we might still have radio contact with to tell them what we had found. As was often the case, a single infantry or rifle Company would operate entirely by ourselves and sometimes a Company like mine would be placed under the operational control of a Battalion different than the one to which we belonged. However, being isolated like we were on this particular mission, we sometimes did not have any contact with Artillery Units or any Reconnaissance Units that might be operating in the area. On this particular night, we were operating entirely alone and had very little contact with anyone. We had just finished digging in and it was almost dark. I was standing talking quietly with Staff Sergeant Lutu, a native Samoan, who was acting as my Company Gunny. We were approximately three feet from each other when something went between us at a tremendous speed, almost knocking us down. Within a couple of seconds, there was an explosion just down over the edge of the ridge where we were dug in.

At almost the same time, there was an explosion in the air above us. Numerous bursts began in the air and over the side of the ridge where we had heard the first explosion. We dove into our holes and thought initially it was "in coming" from the NVA, because it appeared it was either artillery, or rockets, or both.

After hearing the incoming for a minute or so, the sounds indicated it was probably U. S. artillery, because it appeared the projectiles were coming from the east, instead of the west where the mountains led into Laos. We just had to last it out and hope for the best, and the barrage finally stopped after four or five minutes. Fortunately, we did not sustain any casualties.

A short time later, I received a radio call from the 7th Marine Regimental Commander, "bird" or full Colonel Reverdy Hall. I thought it was very unusual, especially since it was not scrambled. He identified himself by his code name and told me he wanted me to move out at first light and do a bomb damage assessment, or "BDA." He went on to say an enemy radio transmission had been intercepted by listening devices on Monkey Mountain near the DaNang Air Base. He added that he had ordered a "battery zone fire" close to my location, and he wanted to know if the artillery fire had inflicted any damages. He also said that he had reason to believe the radio transmission came from a sizable enemy force, and he wanted me to engage them if they were there.

A battery zone fire is having every artillery unit within firing range of the suspected target fire a certain number of rounds into an area about 1,000 yards square. In this case, our problem had been that we had no idea who was doing the firing, and we had no way of notifying them that they were firing on us. Later information indicated that the units firing the mission did not have the correct topographical information and did not realize they were firing over a high ridge. This created a serious problem for us since they were firing both point-detonating and variable-time fuse rounds. As a result, the elevation of the ridge was setting off the variable-time fuse rounds in the air when the sensors indicated they were getting close to the ground, which they were as they came over us. The point detonating-rounds were just clearing us and impacting in the small valley on the west side of the ridge.

I told the Regimental Commander I could do the BDA right then because the rounds had hit over and around my Company. There was a pause, and then he keyed the microphone to speak. Before he said anything, I could hear someone in the background frantically trying to explain to him that we had been hit by our own "friendly fire." He then asked if I had suffered any casualties. I replied in the negative, and then he said, "I don't want to get caught up in any intra murals," and signed off. We did some patrolling the next day, but we did not have any contact. What became apparent was that the radio transmission must have been from an enemy unit with visual contact of us, and they were probably at a higher elevation in the mountains just west of our position.

Interestingly, some years after my release from the military, I read a small book about POW/MIAs written by a former POW. From the maps he had drawn in the book, it appears that on that search and destroy mission we were two-three miles from a POW prison camp where he was being held in the Happy Valley area. That could have explained why the NVA resistance was more defensive than usual there. In trying to compare the timeframes of our mission in the area with what he said was going on in the POW compound at approximately that time, he wrote that they were suddenly moved from that camp west into Laos and forced to walk to North Viet Nam where they were held until the end of the War. I have often wondered if it was because we came so close to the prison camp and maybe the NVA thought we had intelligence indicating that prisoners were being held somewhere in that area.

Following our patrolling, we stayed in the area one more night, and were lifted out the next day to be taken back to 2/7 Headquarters. When the first chopper landed that morning, the crew chief of the CH-46 stepped out of the back of the chopper and waved for us to stay in our holes. A chopper "turned up" or running out in Charlie Country always received lots of attention from the gooks, and usually in a hurry. After a couple of minutes, I crawled out of my hole and went to the front of the chopper where there is a phone in a small compartment on the side of the fuselage that can be used to communicate with the pilots. As I started to ask what was causing the delay, I heard the pilot say, "Hi, Doug." There on the other end of the phone cord was Marine Helicopter Pilot Captain Mike Bryant. We had attended John Brown University together, and neither of us knew the other was in Viet Nam when we first arrived in-country. Mike was the sole member of the JBU Swimming Team back in the world, and he single-handedly placed 6th in the National Swimming Finals in our college classification one year. He was an unbelievable athlete. We obviously did not have any time to converse, but he made up for it later.

"Thou Shalt Not Steal. Thou Shalt Not Kill"

THE SPIRITUAL INFLUENCE of my friends and neighbors in my youth, and the personal values I was taught in and around the small town in Wyoming where I was raised served me well in my military experience. The reinforcement of those spiritual and personal values at John Brown University later as a college student undoubtedly saved my life as I have struggled emotionally through the years. As a society in the Western United States, the foundation of the rural life we enjoyed has been reduced to writing through the years in different ways. One encapsulation is commonly known as the Code of the West. The following version of the Code was written by James P. Owen:

1. Live Each Day With Courage
2. Take Pride In Your Work
3. Always Finish What You Start
4. Do What Has To Be Done
5. Be Tough, But Fair
6. When You Make A Promise, Keep It
7. Ride For The Brand
8. Talk Less and Say More
9. Remember That Some Things Aren't For Sale
10. Know Where To Draw The Line

This Code by Mr. Owens was adopted by the Wyoming Legislature as the Wyoming State Code in 2010. The words have always had a tremendous influence on me personally, but this Code, and the Ten

Commandments in the Bible, have also been a source of agony for me through the years because of my experience with life and death in South Viet Nam. Somehow, I have never been able to reconcile those miraculous teachings of my youth with the excruciating experience of making decisions, some of which were wrong, that led to numerous people on both sides of the conflict being injured, wounded, and/or killed. However, at the time, I reacted based upon my life experience, military training, and what I understood our mission to be. The reasoning behind some of those decisions was not always based upon my rural Wyoming values I had been taught in my youth. In retrospect, maybe I should have been a conscientious objector and avoided a combat role. The depression I have endured through the years brought on by some of the experiences I survived has been made more tolerable because of my prayers that somehow what I did was justifiable in light of the many lives I was responsible for.

Such was the case when I stole things of considerable value. When I took command of Echo Company, I was amazed by the poor quality of the clothing and protective equipment available for my men to wear, and my inability to obtain what were called jungle utilities, or "jungle utes." The standard issue was supposed to be lightweight material with the shirt being worn outside of the trousers. There were large pockets in the shirt and on the legs of the pants. The boots were made of part leather and part canvas material that had grommets in them to let the water drain away from your feet easily. They had steel plates in the soles to help protect against "punji traps" designed to cause sharp spears to penetrate the soles of the boots. The "state-side" boots were entirely leather, and once they were wet, they stayed wet. Sores, blisters, and the ugly infection called immersion foot were common in my Company. Immersion foot was named that because it was caused by a foot being in water continuously.

We were on a search and destroy mission one day and Cathi, who was the best Radio Operator to ever communicate on a radio, told me there was an inbound chopper and we were supposed to quickly clear an area big enough for the Huey helicopter to land. As soon as it landed, Col. Love jumped out and quickly began asking me questions about the operation of my Company. Seeming satisfied with the answers, he asked if there was anything we needed. I pointed to my men and told him we needed jungle boots, and we seemed unable to get any through the supply system. He said he would take care of it and jumped back in the chopper. I noticed that another 2/7 Company Commander was sitting in the chopper as they lifted off. The chopper returned in just a few minutes, and the crew threw out 15-20 pairs of what seemed to be new jungle boots. We found out later that as soon as Col. Love landed at the Battalion Headquarters, he told everyone there who were wearing jungle boots to remove them. He threw them in the chopper and sent them out to me. We also found out later that the Company Commander who was in the chopper had just been relieved of his command prior to Col. Love coming to my Company. He was supposedly relieved because he could not satisfactorily answer the Battalion Commander's questions about the operation of his unit.

The boots Col. Love sent me were a "Godsend," but the problem with the jungle "utes" still existed: I had men with jungle utes that were literally rotting off their bodies. Their bare skin was exposed to all of the elements, and when we moved to Hill 190, we were operating much more of the time in the jungles, instead of the paddy areas. It was a situation I believed was intolerable.

Because we had provided security for the Navy Seabees on Highway One up and over Hai Van Pass, I had been to the Red Beach Supply Center north of DaNang where the Seabees Headquarters was located. It was a complex of unimaginable size, and the enormous amount of supplies was for much of the I Corps area, which was about 20% of South Viet Nam. 2/7 would send us a Six-by truck to use occasionally, if I could convince our Battalion Supply Officer we needed it for some particular reason. I do not remember what lie I told the Supply Officer, but I ended up at the Red Beach Supply Center in a Six-by with three of my Marines and the truck driver.

We were able to bluff our way in the gate because of my rank, and we drove around through the massive crates and equipment until we saw what we thought were crates of jungle utes, because of the markings. I told the driver to park next to the crates which were about 10-12 feet high. We boosted two of my Marines up on top and I told them to break open the crate with a pry-bar we had found somewhere. When they had it open, we discovered we had found the right crates. They immediately began throwing armloads of jungle utes into the back of the truck. I was afraid we might lose some of them on the way back if we took too many of them, so I told my men we had enough and to do the "snow angel" spread position on top of this huge pile of clothing that was approximately three feet deep. The truck driver was horrified, and when I told him to head for the gate, he said he was going to have to stop, and that he did not know what to tell the guard at the gate. I told him not to stop, and he told me again that he had to. Looking him in the eye, I calmly told him that if he stopped, I was going to shoot him, dump him out at the gate, and that I would drive the truck out of the Red Beach facility.

He apparently believed me, because when we got to the gate, he accelerated the truck and we disappeared in a cloud of sand and dust. It turned out that we had over 60 pairs of jungle utes...and a court martial offense for the person who carried out this "grand theft" if he were ever caught. When the driver left Hill 190 later that day, I told him if what he participated in was ever discovered and he was going to be prosecuted, I would be there to tell the truth and take responsibility for my actions. I never did hear from him again.

The shirt part of the jungle ute we wore was officially named a "blouse." The material offered very little comfort in the cold rains during the monsoon season because it was very light, but amazingly strong and durable. However, the jungle utes dried quickly when they became wet, which they often were. In addition, we wore our helmets, cushion-soled socks that were "calf high," and jungle boots...but no underwear. We all tried to keep an extra pair of socks with us, and when a wet pair wasn't being worn, we tied that pair around our abdomen. We had discovered that just body heat would help dry the wet socks much more quickly than could be imagined.

One of the missions at Hill 190 was the use of a huge, powerful spotlight to harass and interdict possible enemy movements at night in and around the villages and hamlets in my TAOR. We would start the generator at different times during the night and light up areas where there might be enemy movements. I always assumed that it was not very effective because the sound of the engine starting always preceded the light coming on, but it was one of our missions.

Our generator began to fail, and I could not get any response from the Battalion as to how I could have it repaired. When it appeared that there was going to be a catastrophic engine failure, I told Jack to order some Marine green paint. It was to be in spray cans, and as he always did, he had it sooner rather than later. I asked for another truck to haul supplies, but by the time our supplies were delivered back to our Company location on the day of the delivery, I made sure it was near dusk.

Travel at night in a vehicle was a "suicide mission," especially when you had to go through five or six villages and hamlets and often, as mentioned previously, we would find mines buried in the roads in the mornings. On this night, I reported to Battalion Headquarters that it was too late for the truck to return to the Battalion area, and that it would remain at our Company location until morning. After I sent the message, the driver and I headed for Red Beach. Cathi and his radio were with us. The driver knew what we were going to do in advance, and he had agreed to volunteer for "the mission." I did not take anyone other than Cathi with me because of our need for patrols, ambushes, and protecting Hill 190. Additionally, I did not want anyone else exposed to the potential consequences if what I was going to do was discovered.

Again, we did not have any problem entering the Red Beach supply depot. We found the large, wheel-mounted generators fairly quickly, but they were all painted Air Force blue. I would guess the generators weighed 1,500-2,000 pounds. I had a piece of scrap paper and a piece of cardboard shaped like a clipboard and acted like I was checking the serial numbers. It was not difficult to find one that I liked, but when the driver backed up to the generator, Cathi and I could not lift the tongue of the generator to hitch it to the truck. Playing my "hot-shot officer" role, I was able to get a forklift driver to "hurry up and help us." After we were "ready to roll," we headed for the gate.

It was an exceptionally dark night, and there were few lights in the facility. They used as few lights as possible, because the more lights that were on, the easier it was for the "bad guys" to be able to aim at the facility with rockets from a distance. When the guard stepped out to stop us, he nearly got run over, and all he saw in the darkness was unrecognizable. A truck of that size traveling at night with little or no lights, depending on where we were, should have attracted a lot of attention. Thankfully, as the old saying goes, "The Lord takes care of fools and children," but that driver, Cathi, and I were not children. The poor people in those little villages and hamlets must have thought a major attack was beginning when we roared through them in the darkness on the way back to Hill 190.

By first light, the new generator was painted Marine Corps Green, and the serial numbers had been filed off. The generator was operational the next night after "it was delivered," and we were once again able to accomplish our mission. The driver took the truck back to the Battalion Headquarters after having to "stay at Hill 190 all night," and once again, I held my breath for a few days.

I did not allow my troops to have cash, because it could be used to buy drugs in the villages. As a consequence, they had to wait to draw their pay when they went on R&R. There was nothing you could buy where we were...except drugs. The NVA and VC would come into the villages with packets of 25 and 50 joints of pot and force the indigenous people to take them. They would tell the local Vietnamese they must have the money for the drugs in American "greenback" currency within a certain time limit, or they would return and kill all the old men...or kill the children...or burn down their hooches. With those kinds of menacing threats hanging over their heads, the villagers were desperate to sell the drugs to us when we moved through the countryside, villages, and hamlets. Because of my prohibition of available cash, some of my men might have traded their C-rations for some dope, but I believe it was seldom because we had enough trouble getting food as it was. I tried to help the Vietnamese people in the best ways that I could. I carried 7,500-10,000 Vietnamese Piasters with me all the time and I would buy the drugs from the villagers to assist them, even though I never paid them with greenbacks. Jack would always destroy the drugs for me. That was the best I could do.

Another problem we had was getting 782 gear. 782 gear consisted of helmets, helmet liners, flak jackets, cartridge belts, and water canteens, among other things. Jack, Sid and I came up with a plan to get some of those things because, once again, we could not get them through the supply system. Even though I did not allow anyone to have cash in the Company, I made arrangements for Sid and me to draw $20 each. We assumed that when Marines were killed and sent to the morgue at the DaNang Air Base, their 782 gear was removed before their remains were sent home.

With that in mind, we decided to have Jack get a ride to DaNang and find out if what we thought were true. We gave him the money and he caught a ride to the DaNang Air Base. He discovered that our assumptions were true, so he went to the Air Force PX and had a U.S. Airman buy him three cases of beer with our money.

He then "borrowed" an Air Force jeep that was parked outside the PX and he took the beer to the morgue. The Gunnery Sergeant in charge of the morgue agreed to take the beer for a pile of the 782 gear since, by government regulation, it was only being destroyed. Other than the partial remains of the deceased Marines or other military personnel that occasionally might still be on the gear, it was very usable and provided safety for my men. Without the gear, it could have led to the injury or death of my men. Jack returned driving his borrowed vehicle with the 782 gear we desperately needed.

Our plan also worked a second time, but the third time Jack went to the morgue in our borrowed Air Force jeep, a new Gunnery Sergeant was in charge, and he had Jack arrested. I received a message from Col. Love that one of my Sergeants was in the Brig at the DaNang Air Base, and he wanted to know, "What the hell is going on?" (Expletives deleted!) I told Col. Love that Jack had "disappeared" the day before and that I would go get him.

Col. Love sent a jeep out to Hill 190 to pick me up and we went to the morgue at the DaNang Air Base. On the way through the Base, I realized that I was in the wrong branch of the service when I saw American military personnel playing softball games with teams wearing uniforms...and no one was even carrying weapons. They looked at me like I was from outer space, and I later thought I should have been in outer space when I walked into the comfortable, clean office of a Colonel who wondered what kind of an idiot I was. He demanded that I explain how I had allowed one of my men to be in DaNang instead of being with his unit.

I agreed with everything the "good Colonel" said, and I assured him that this Marine of mine was going to be court martialed. In addition, I adamantly assured the Colonel that my AWOL trooper would spend the rest of his tour in Viet Nam "on the point" of every Operation. I think I even used words the Colonel had never heard before when I described what was going to happen to Jack. After a few minutes, they brought Jack in and I "really threw a fit." Jack looked at me like I had lost my mind, but it only took him a few seconds to realize what my "gig" was. They finally released him to me, and we sped out of there in another cloud of dust. Needless to say, that ended our ability to get 782 gear that was sorely needed, even though it was just being destroyed to satisfy some government regulation. As you might guess, we also lost Jack's borrowed Air Force jeep.

One of the problems I had to deal with personally was the killing of other human beings. Did the fact that some of those human beings were enemies who were trying to kill us have meaning? For me, it did.

However, in jungle warfare where ambushes were an obvious tactic, and the killing was "up close and personal," I did have a revulsion when we killed combatants who, at that particular instant, did not have a chance. The words of General Patton I referenced earlier would "ring in my head" each time I doubted the necessity of killing them before they had the chance to kill us. However, my spiritual values could only be "masked" in my mind; those values never changed, and it still haunts me to this day. My apprehensions could be best expressed by Cathi's description of an enemy encounter in his personal writing titled *Shaped by the Shadow of War* (pg. 384-386):

> The day after I killed the bamboo viper, we were walking along one of these large trails and we heard voices and smelled food cooking. The Skipper and I and a half dozen Marines quietly edged up the trail. We paused there for a long minute just listening to them. It sounded like there was a group of about half a dozen Vietnamese--which meant NVA soldiers--eating breakfast in a campsite off to the side of the trail. They were inside a hedge that made it hard to sneak into a position where we could capture them. They were obviously unaware that we were on the other side of the hedge, only about 10 to 12 feet away from them.
>
> Captain Chamberlain leaned over and whispered, "I want one more fire team up here. Tell everyone else to stay put and stay very quiet."
>
> I nodded and waddled back up the trail a little ways so that I could talk over the radio without making any noise that the NVA could hear. I put my hand over the mouthpiece of the radio handset and whispered into the radio. "Echo Two, this is Echo Six, Over."
>
> Echo Two Actual came on the radio (Echo Two was the radioman. Two Actual was the Lieutenant.) "This is Echo Two Actual, over."

"Two Actual, this is Echo Six. We've got a group of gooks right off the trail up here. Skipper wants you to send one more fire team up. Tell them to be very quiet about it." "Roger, Six. Will do." "Six out." A moment later we had four more Marines. Captain Chamberlain motioned for everyone to form a semicircle around the area where the enemy soldiers were. He gave me a grim look and shrugged. I realized his shrug meant he was saying that there was no sense trying to crawl into this hedge and capture these guys. Then he pointed his rifle toward the thick hedge, and everyone knew what we were supposed to do.

When everyone was in place, we opened up on them with our M16s—firing blindly through the hedge. I kept my fire close to the ground and sprayed back and forth as I fired off the magazine. We quickly replaced our empty magazines and went into the campsite. They never had an opportunity to fire back.

One of them was not killed outright; we followed his blood trail and found him a short distance away. Several Marines gathered around and stood over this guy, who was still breathing. He lay there looking up at us as he died. Then I was surprised to see the corpsman get down on his knees and try to revive this guy with mouth to mouth resuscitation. He was really working hard at it too.

It did feel mighty strange to see the corpsman try so hard to save a guy that we had just made every effort to kill. I felt sympathy for the corpsmen; they were forever trying to keep dying guys alive, often an impossible job. Our job as grunts--just killing the enemy--was a lot easier. The corpsmen had a completely different orientation, and they definitely had the tougher job. Killing people was nothing compared to the pain of trying to stop people from dying.

The Military Industrial Complex

UNITED STATES ARMY FIVE-STAR General Dwight D. Eisenhower became the 34th President of the United States after his retirement from active military service following the successful conclusion of World War II. He publicly expressed his profound concern about the relationship between military research and development, or R&D, and the resulting procurement of weaponry, equipment, vehicles, and supplies as well as other expenditures recommended by the military leaders of our Country. The purchases were being made from the growing industrial complex that had been developed to meet the enormous needs of the military effort to win WWII.

As a First Lieutenant and a Captain, I was not privy to any of the complex issues that affected government policy, but at the Infantry Company level, we had to deal with the results of those major policy and procurement issues on a daily basis. I understood that R&D may have its limits in peacetime, but it appeared to me that some of the R&D in our era had not been pursued to a satisfactory point, and that a military confrontation was apparently needed to further develop and refine the weaponry.

In my opinion, a tragic example of that was the M-16 rifle. As I mentioned earlier, in 1965-66 the Marine Corps was still using the M-14 rifle to train Officers in marksmanship. I never saw a M-14 used in Viet Nam except to shoot smoke grenades for marking positions in the jungle canopy because the design of the M-16 prevented it from shooting a smoke grenade the necessary distance that was needed in the jungle environment. The M-16 had every appearance of a weapon that should be a good piece for use in the "close-quarter" combat of a jungle environment because it had the capability of firing the 750 rounds per minute. However, a 20-round magazine was expended so quickly that the initial burst usually had to be followed by the removal of the magazine and the insertion of another.

When you are engaged in close proximity with an enemy that has an AK-47 with a 30-round "banana clip" that is flawless, it is obvious who has the advantage, even though it may fire a little slower. The spring in the magazine for the M-16 that pushed the rounds upward as the weapon fired and ejected the previous round had a huge design problem. The spring was so weak that it often could not push the rounds up with enough pressure to adequately sustain firing in the automatic setting. As a result, we could only put 18 rounds in each magazine and each round had to be loaded manually, one at a time. As a consequence, the loading device that was supposed to allow the magazine to be quickly reloaded with 20 rounds was, for all intents and purposes, useless.

When we could get tape of some kind, we would tape two magazines together so one end was inserted, and the other was pointed downward. The empty magazine could then be removed, and the other loaded magazine could be inserted quickly...if everything worked right.

A weapon that can fire as fast as the M-16 must be of a quality that is immune to dirt, dust, gun powder, water, heat, and several other things that could possibly cause it to malfunction. The tolerance of the moving parts in the M-16 was so close that almost anything would cause it to jam. The major mechanical malfunction was caused by the slightest contamination in the chamber, even powder residue, which could impede the ejection of the expended round. In a weapon that fired that rapidly, the slightest resistance to the round being removed from the chamber would often cause the ejector ring to be broken off the expended round in the chamber. Then, another round would be instantly chambered into the shell casing that was stuck in the chamber, and all of it would happen in "less than the blink of an eye."

To clear the jam, it was necessary to get the cleaning rod out of our packs, assemble it, and "hammer" it down the barrel until we were able to knock the chambered round out of the shell casing in the chamber. Then, we had to have an extractor to remove the shell casing that was still stuck in the chamber of the weapon...and all of this while we were engaged with an enemy who was trying to kill us. Another issue was that while the velocity of the M-16 round was so fast that it created immense tissue damage when it struck a human body, those same rounds would not penetrate the thick jungle foliage like the AK-47 carried by the NVA and VC.

Possibly the most critical issue was that every M-16 had matched parts, meaning that the parts were not interchangeable and could only be used together as one set of parts. The AK-47 parts were interchangeable. The M-16 would field strip into seven components while the AK-47 only contained five parts. In addition, the Chinese/ Russian weapon had a folding bayonet that was secure on the barrel of the weapon and it could be folded into position simply by pressing a button. To make it even more lethal, the folding bayonet was pointed on the end and gradually expanded into three cutting edges as it extended back toward the barrel.

There was no comparison between the weapons. An AK-47 could be placed in mud and water and it would function properly as soon as the trigger was pulled. The M-16 was shorter in length and did not weigh as much as the AK-47, but the M-16 was so dysfunctional that during the period of time I was in command of Echo Company, our rifles were replaced twice with a newer version of the M-16 rifle in an attempt to get ones into the hands of our troops that were reliable and would function properly.

The M-16 reliability issue was critical, and then, unbelievably, we were told that we were being rationed ammunition on a monthly basis. That had to be one of the most ignorant decisions ever made in a combat situation, in my opinion. As you might imagine, I immediately wondered what we were going to do for the rest of a month if, on the 20th of the month, we ran out of ammunition.

I first worked for someone when I was nine years old near our town in southeast Wyoming. I mowed and raked hay driving small Fordson and BN International tractors. There were not any child labor laws then, and the farmer I worked for was an innovative agricultural leader named Curt Meier. Among the many things he taught me was a simple truism: "Don't worry about things you can't change." With that in mind, I made the decision to "sling" our M-16 rifles that did not work and for which ammunition was being rationed, and to start using captured weapons. "Sling" meant we just carried the M-16s on our packs. We had access to captured AK-47s, M-1 and M-2 carbines, Thompson sub-machineguns, pistols, and a good supply of captured ammunition that we could obtain on a "regular" basis.

One day on a search and destroy mission, Jack was carrying a "like new" M-2 carbine with a folding steel stock that we had captured. The M-2 had a selector switch which allowed it to be fired either automatic or semiautomatic. Jack, even though he was a Sergeant, was a Platoon Commander that day and I had him and his platoon deployed on our right flank next to a road. I noticed a jeep come down the road and stop close to where I thought Jack was. Later that day when we were digging in for the night, Jack came to me and told me there were two Marine Lieutenant Colonels in the jeep. He said they stopped when they noticed he was carrying the M-2 carbine, which was not an authorized weapon for Marines. They disarmed him and took the weapon with them. As if on key, a few minutes later I received a radio message from Col. Love that a chopper would pick me up at first light the next morning and take me to Battalion Headquarters.

The chopper arrived very early, and after the short ride to 2/7, I went to Col. Love's hooch. He told me he had received a complaint from our Regimental Commander about one of his Companies carrying unauthorized weapons and he wanted to know what the "!#%@$&!!" was going on. He and I had shared numerous comments about the unreliability of the M-16 in the past, so he knew the basis of the problem. I told him as long as I had a mission to accomplish, I was going to carry it out to the best of my ability by providing for the safety of my men. I told him we were using weapons that were reliable and functional, and for which we could obtain our own ammunition. He paused for a few seconds, and then said, "Alright you dumb S.O.B, but don't get caught at it again!" I returned to my Company, and we continued to carry our captured weapons..."discreetly."

Marine Corps issue to infantry units was the M-60 machinegun, which was an improved model of the ones used in World War II and Korea. The older Korean-era models that had been developed had a "water jacket" to cool the barrel during sustained firing. That had been proven to be impractical, so the newer M-60 barrel was air cooled. However, a major problem with the M-60 was that it seemed to invariably misfire when it was used to trigger an ambush. Without elaboration, it is a serious problem when you hear a click instead of a burst of rounds when you are trying to surprise and kill the enemy in an ambush. The corrosion of the rounds in the jungle environment was also an ongoing problem.

Partially as a result of the problems with the M-60, we carried Winchester 12-gauge pump shotguns. We used magnum flechette rounds that had the explosive power to force the small, pointed, "tack-sized" steel projectile with a vaned stabilizer tail through one inch of plywood at 50 yards. They were excellent, reliable, and deadly. We used them to trigger ambushes with great success. Someone once said that shotguns were prohibited by the Geneva Convention on the Rules of War. If that is true, we violated that rule. Flechette rounds were also developed for use by tanks and artillery units, and a "firecracker" anti-personnel round had also been developed for artillery missions. When the firecracker round hit the ground, it would pop open scattering projectiles similar to small hand grenades. The small projectiles would then all start exploding. It was believed that these kinds of munitions could do more damage to enemy personnel than the single blast of a traditional round.

The expanded use of napalm brought about horrific results as far as collateral damage to civilians. The resilience of humans is amazing, and we saw the indigenous population use banana tree leaves to wrap around the burned portions of their bodies because the oil in the leaves would soothe the burning from the napalm and help in the healing process. Undoubtedly, napalm is one of the most horrific and disfiguring substances ever used against human beings.

These are just a few examples of the development of weapons for use in war, and whether the use of them was considered to have been correct or not, they generated tremendous profits for "the Complex."

Chain of Command

As I mentioned previously, the Marine Corps had numerous officers who came into the Corps after the Korean War, and they had too much rank to be Platoon Commanders or Company Commanders in the Viet Nam War. As a result, they did not have the opportunity to lead combat units in South Viet Nam. However, they were in positions to make decisions that affected combat units without them having had any actual experience in combat situations. Some of them tried to "micromanage" units they had contact with or some kind of tactical control over. 2/7 was no exception.

I had to acknowledge the organizational requirements of the Battalion and work with a few officers whom I thought were not qualified to be making decisions because of a lack of experience, just as some of my troops initially felt about me. The difference was, they did not have any Noncommissioned Officers under their supervision to assist them like I had. Rather than acknowledging that they were restricted in their knowledge, some of them were very arrogant and overbearing.

When I thought the safety of my men was being jeopardized, I would take whatever action I thought was necessary to protect them. I had a burning desire to never compromise the integrity of my unit for the gratification of others or to do things I thought did not have a useful, tactical purpose, just to be following orders.

Determining ambush sites was done through the observations of trails, stream crossings, and areas of good concealment where there appeared to be considerable foot traffic by humans and animals. Each patrol or movement by anyone in my Company was "debriefed" by squad leaders, platoon leaders, platoon commanders, or Sid and me, as much as possible. It was informal in a manner like, "See anything?" or "What did it look like?" or "Anything we can use out there?" That information would be given to me each day for Cathi and me to use when determining possible ambush locations and "H&I" targets of opportunity that night. H&I meant Harassing and Interdictory artillery fire.

The H&I process was for me, as a Company Commander, to send artillery units in my area of operation a list of map coordinates each day of locations where there might be targets of opportunity. If the available artillery units had a lull in activity at night, they would fire a number of rounds at some of those map coordinates randomly after dark. While it may seem that it was a "crapshoot" with little chance for success, VC combatants and NVA soldiers we captured indicated that those artillery rounds suddenly hitting locations at night without any kind of warning were very psychologically damaging and demoralizing to them. The bottom line was that I had pertinent information on a daily basis, and I made tactical decisions concerning ambush locations and H&I targets of opportunity based upon real-time intelligence. This was in stark contrast to what appeared to be a good idea someone had who was looking at a map while sitting at a desk some 3-10 miles away from my Company location.

An example of a serious problem that developed was when we were patrolling and setting ambushes during the monsoon season. The policy in our Battalion, and possibly the entire I Corps Area, was that troops involved in ambushes would "break bus site," or leave the ambush location, at 0100 hours, and move back to the Company perimeter. This requirement was not to be followed occasionally. It was to be adhered to every night, which I thought created a predictable pattern for the enemy to exploit, but that was the strategic policy. Linking up armed units in the dark of night was very risky. The use of flares to signal each other when we needed to made it easier, but it also compromised everyone's position. The ambushes were required to do radio checks every hour, and it was done by Cathi while he was there. Each hour after the ambushes were set, the ambush number was whispered by my Radio

Operator, and the ambush Radio Operator would click his handset once if everything at the ambush site was O.K. Then at the given time of 0100 hours, the Radio Operator with me would whisper "break bus site" and the ambush was supposed to start moving. Every night, each ambush had checkpoints to whisper to my Radio Operator as they moved back to the Company location.

My Radio Operator would keep track of the progress of the ambushes on a plastic-covered map with the use of a crayon-type marker and the use of a flashlight with a red lens. The Operations Officer, or S-3, for our Battalion had all of our ambush sites on his map at the 2/7 Headquarters. He required that we send them to him every afternoon prior to our ambushes being set that night. Invariably, the S-3 would tell us that some of our ambushes we planned were in the wrong places and he would order us to change them. I tried to accommodate him and follow his orders, but as the monsoon season came on, it became more difficult. He would make us change locations of ambushes and place them in areas covered with water, and he would not accept my reasons for not agreeing to his changes.

Ultimately, we were "pushed into a corner" where we had to acknowledge what he said to do, and then keep a separate set of ambushes that we were actually using. Cathi would have to call in the "phony" ambush information all night to the overbearing S-3, while at the same time, keeping track of the actual information concerning our real-time ambushes. It all "came to a head" one night when Cathi was exhausted, partly because of the hassle of keeping two sets of ambushes every night. The S-3, a Major, sat in his air-conditioned hooch miles away and apparently watched with some sort of personal satisfaction as Cathi erred and indicated that two of the ambushes had walked into each other in the darkness after breaking bus site. After the information Cathi sent indicated that the ambushes would have more than likely opened fire on each other in the darkness if they had been where our reports indicated they were, the S-3 sent me a radio message and asked me how many casualties we had from our ambushes running into each other. It was obvious he was willing to have my men wounded or killed just to make a point rather than warn us or ask questions about what appeared to be happening in the intense darkness.

Once again, Col. Love had a helicopter pick me up the next morning. When I walked into his hooch, the S-3 was standing near Col. Love's small table with a smirk on his face. When Col. Love asked me what had transpired, I gave him the history of what I considered to be interference by the S-3 in the operation of my Company, including his indifference to the safety of my men "simply to make a point." I went on to say that it was imperative for the Commander in the field to make the decisions he felt necessary to accomplish the mission that was assigned to his unit. Col. Love did not even hesitate. He told the S-3 to never interfere with the operation of any Company in the Battalion again and that if the S-3 had questions about a particular unit, he was to discuss it with him, meaning Col. Love. He then thanked me and told me to "Get back to work."

As an aside, the flares we used that were so crucial to communicate while trying to link up in the darkness were in very short supply. We often had to trade our C-rations to South Vietnamese military units to get flares they had because we could not get them through our own supply system.

Propaganda

IT IS DIFFICULT TO FULLY UNDERSTAND the effects of propaganda in a war, and even more difficult for the general public to completely identify it. We became aware of it and the activity it can generate with the incident involving the professor from Ohio University. Information sent to us by relatives concerning the anti-war demonstrations was shocking, and while it was considered free speech back in the world, to us it was demoralizing, and it caused immense feelings of foreboding and betrayal. It was not unusual for us to find leaflets written in very crude English illustrating what was going on in America, how the American people did not support us, and that this was a war perpetrated by the Johnson Administration and the United States military leaders. The propaganda included overt attempts to inject racism into the efforts to affect us psychologically.

One incident happened during the weeks leading up to the 1968 Tet Offensive. A few miles north of DaNang on Highway One was a small village named Namo. It was located on the south side of the Song Cu De River near the Gulf of DaNang. As you might expect, there was a bridge there for use by the traffic on Highway One: referred to as the Namo Bridge. The bridge had been destroyed by NVA Sapper Units prior to my arrival in-country, and a pontoon bridge had been constructed by the U.S. Navy Seabees for temporary use to keep vehicle traffic moving over Hai Van Pass north to Phu Bai and Hue. Marine Military Police Units provided traffic control at the bridge.

As the MPs were arriving one morning, they were ambushed as they entered the village. The jeep they were traveling in was hit in the radiator with a B-40 rocket, wounding all four of the MPs. The VC laid the wounded Marines on the ground and gave medical attention to the only black member of the four-man team. They placed hand grenades on the chests of the three white team members and "pulled the pins" causing the grenades to explode ultimately killing those three Marines. The black survivor was given written pamphlets to take back to his unit that were full of inflammatory, racist statements such as, "Black man, do not fight the white man's war." and "The white men are raping your wives in America while you are here killing innocent people." and "We are killing the Marine dogs and we will kill you." Those were easy to identify, but other efforts to demoralize us were more covert.

In the late fall of 1967, the Johnson Administration suspended the bombing of part of the Ho Chi Minh Trail. Was it propaganda generated by our own government indicating that we were willing to take actions that could lead to peace, or was it an attempt to allow the NVA/VC time to reinforce their positions and supplies so the war could be extended for the economic benefit of the United States? This will be discussed in more detail. (See **Tet** chapter.)

I will admit that I do not feel the guilt now that I felt in those days for thinking we were being sacrificed for the economic interests of our Country, especially after reading excerpts from the book written by then Secretary of Defense Robert McNamara. I now believe Secretary McNamara and President Johnson were traitors after reading their acknowledgement that they decided very early in the war that it could not be won in the manner they were conducting it, and they made no plans to change the conduct of the war. I will elaborate more about my accusations in a later chapter.

The Hanoi Hannah broadcasts from North Viet Nam were a venue for anti-war advocates like Jane Fonda, who referred to American military personnel as "pigs" and "baby killers." The actions of Jane Fonda can only be described in a manner not suitable to be included in this book. The small transistor radios that had been developed in those years were very accessible to military troops. The radios were not only one of the main sources of the demoralizing anti-war rhetoric reaching my troops, they were also an ongoing threat to our safety because of the understandable desire of young Marines wanting to have some kind of contact with the world.

Winning the Hearts and Minds of the People

THE SOUTH VIETNAMESE SOCIETY in the countryside was divided by two religious beliefs...the Buddhists and the Catholics. They lived in separate villages and hamlets. The contrast was amazing to me. While both groups lived as people did hundreds or maybe thousands of years ago, our relationship and interaction with them was dramatically different.

The Buddhists were pacifists and were literally slaughtered by the Communist NVA and Viet Cong if they did not do exactly as they were told. Part of our efforts had to be to protect this segment of the society from this violent enemy by utilizing violent means, while the people we were trying to protect did not believe in our method of protecting them. They clung to their ancient beliefs and wanted simply to be left alone. They would do whatever they thought they could to meet the demands of whichever group was in control of their area at any point in time, given their religious beliefs.

The Buddhists seemed to celebrate something every week or so. When they did, they would invite "Honcho," which was me, to the festivities. They would crowd around me while we were eating, and the children were so fascinated by the hair on my arms and hands that they would keep touching and feeling my skin. After the first celebration, I thought Sid would make a much better representative of the U. S. Government at those events than I did, and several times I gave him the opportunity to prove that my contention was true.

The Catholics, on the other hand, had their own weapons and did everything they could do to protect themselves, including cooperating with us as much as possible. They were tenacious fighters who continuously resisted the overwhelming violent threats of the murderous VC and NVA.

The one commonality both groups had was that they methodically tried to farm their plots of land in the same manner they had since ancient times with the use of water buffaloes, hand labor, and human waste as fertilizer. Rice and raw fish were the staples of their diet, beef was a delicacy, and raisin wine was a celebratory drink.

One overwhelming problem of working with the indigenous people was that very few of us, if any, spoke or understood the Vietnamese language. We were at the mercy of what were called Kit Carson Scouts who worked as interpreters, and we had not received any training or education in the language prior to our arrival in-country. It was a common belief that these Scouts had been infiltrated by the Viet Cong, even though they were provided to us by the South Vietnamese Government. They were somewhat fluent in English, but the communication with them and the Vietnamese people was difficult at best. However, the Scouts did have a profit motive to "kill the enemy." The South Vietnamese government paid them the equivalent of $20 for every set of human ears they turned in. My problem was that I wondered how the South Vietnamese Government could tell whose ears they really were, and it seemed to me that some of those Scouts would have been happy to sell them ours, especially mine.

A problem with the severing and selling of ears was that under our military regulations, it was considered mutilation of a body, dead or alive. These Scouts considered my attempt to enforce our military regulations as interfering with their livelihoods to support their families. One incident happened when we were near the coast in an area near Phu Bai where there were some small sand dunes. As we moved along on a search and destroy mission, one of the Scouts suddenly began to shout, "VC, VC!" None of us had seen anything up to that point that would make us think there were any enemy troops in the area. The Scout then fired a M-16 burst into the ground just to my right and behind me. As my men close to me tried to react to what appeared to be enemy contact, we suddenly could see that the Scout had killed a VC in a spider trap that was covered by a bush.

The Scout then quickly jumped down beside the dead VC and began to cut his ears off. I shouted at him to stop, and he paused and glared at me with apparent hatred and disdain. As we stared at each other for a few seconds, many issues raced through my mind. I was fully aware of the U. S. government policy concerning mutilation. However, I was also aware that this Scout was attached to us and was, for all intents and purposes, inside our lines continuously. I did not want someone inside our unit who would have a reason to dislike us any more than they possibly already did, because the danger was obvious and many of my men could have suffered the consequences. In those few seconds, I decided I needed to continue on with the operation of the Company and leave the issue of payment for pairs of human ears to someone who had a higher pay grade than me. In addition, the Scout probably prevented the gook in the spider trap from opening fire on us from behind. The VC were very astute and knew the general operational methods of our infantry units. This VC was located inside of our formation and may have been trying to determine who I was in order to take out the command personnel, which they often tried to do. Whatever he was trying to do, he failed...and I assume the Scout was paid for his pair of ears.

Mutilation of bodies in a war is a terrible thing, but it is very evident that it has a dramatic effect on an enemy. Either it enrages an opponent to retaliate in even more barbaric ways, or it "puts the fear of God in them," and they fear the consequences of their actions. We had received unofficial reports that a Marine unit had been overrun and the Marine's penises were severed and placed in their mouths. In another instance, when I had gone to 2/7 for my promotion, I was told by intelligence personnel that a Marine had been captured by enemy forces near DaNang, and his hands were tied behind his back. A bag with two rats in it was supposedly placed over his head, and the bag was tied with a rope around his neck below his head. According to the report, the rats consumed the flesh from his head, including his eyes, and he was still alive when he was found. These sorts of accounts, even if not known to be from a reliable source, cause a mixed sense of fear and hatred in a warrior.

On another occasion, a VC was able to be in a position inside of our unit as we moved on a search and destroy mission. He jumped up from a spider trap just in front of me, and he opened fire on me with his AK-47 equipped with his 30-round banana clip. He was so close to me that I could feel the muzzle blast on my face. Somehow, he missed me. Because he was inside of the Company formation, no one could easily return fire because of the danger of hitting our troops who were dispersed all around the area. In an instant, which seemed like an eternity, he was gone and disappeared into a tunnel. We threw a couple of grenades into the tunnel, but I did not take time to pursue him. We had learned from previous experiences that part of the reason the gooks tried that style of ambush was to get the unit to stop, which could have led to even more complications. A few gooks in tunnels could inflict considerable damage to a Company if they were able to decoy the movement of the unit.

When we were "digging" in that night, I dropped my cartridge belt that I wore around my waist to make it easier to dig. I noticed something that gave me pause. A bullet had grazed my belt so close to my abdomen it tore through the heavy material, but again I had been spared...and once again I thanked God for His mercy.

Quite frankly, actions that cause alarm because of deeply held religious beliefs can also be very dramatic and effective. We had moved north to the Phu Bai area shortly before the 1968 Tet Offensive, and the first night in the area we encountered what later appeared to be Chinese soldiers. One of my four-man listening posts was attacked by a force of ten or twelve enemy combatants. While my troops were wounded, they were able to kill and wound five of the attackers. The enemy bodies we recovered were as tall as six feet and did not resemble the Viet Cong or North Vietnamese we had previously encountered during my tenure of command of Echo 2/7.

My assumption was that they were Chinese, which intelligence sources had been reporting were allied with the VC and NVA. My men said they appeared to physically be under the influence of something because even after they were shot several times they appeared to continue to advance before finally succumbing to their injuries. We had not seen this before. One of the enemies that was killed had taken an entire magazine of M-16 rounds in the middle of his chest in an area smaller than a fist. It was possible to see completely through his body because of the hole in his chest.

The next morning, I approached the people in that Buddhist village and asked for information about NVA and VC activity in the area. They refused to talk to me, so I did what I thought was in the best interests of the safety of my troops and to complete our mission. There was a small footbridge close to the village that was the only way for the civilian population to cross an old irrigation canal dug by the French. It was full of water. Without the bridge, villagers would have to travel a long distance to get across the canal.

Knowing some things about the superstitions and religious beliefs of the Buddhists from a comparative religions class I had taken while attending John Brown University, I had my troops place the dead "Chinese" with the hole in his chest on the end of the footbridge in a sitting position with a lit cigarette in his lips and a C-ration can in his hand. We also stuck a lit cigarette in the hole in his chest to make the situation even more "gross." The villagers were terrified by the dead body and would not go near the footbridge. It became a waiting game. Eventually, after several hours, the village elders came to me with information about what was going on in the area as far as enemy activity was concerned. In return, I allowed them to move the dead body so they could use the bridge. However, the disposal of dead bodies was left up to them so they could accomplish it within the provisions of their religious beliefs.

In another encounter in another location, I had moved my Company into a position outside a hamlet after dark and we dug in. It was slightly higher than the surrounding rice paddies and offered us excellent fields of fire in case of an attack during the night. As the early morning light began to illuminate our position, I discovered that I had unknowingly moved my Company onto a burial ground, and the damage to this sacred site was very significant. The indigenous people in the area were beyond furious. Nothing I could say or do could repair the damage we had done to the burial site and the perceived disrespect they believed I had shown toward them, even though it was by mistake.

There were also social problems that occasionally resembled issues in rural America. The rice paddy areas were free fire zones at night, and no one was permitted to be outside of their villages from darkness until dawn. One night, one of my men killed what we thought was an enemy combatant in a rice paddy with the use of a "starlight scope." The starlight scope was a device that could somehow intensify even the smallest amount of light, like that generated by stars, and it would illuminate the terrain for a variable distance determined by the objects in the proximity of the scope. The scope was mounted on an M-16 rifle, but we only had one or two of them and they did not always work. They were the precursors of the modern night-vision equipment used by military units today. It was an amazing shot of about 200 yards, which previously was thought to be almost impossible.

When we went to where the body was lying at first light, we discovered it was a man who was a local farmer. After looking at the location where the killing took place, the local Village Chief said that the farmer had gone out there and dug a hole in the paddy dike to steal water from his neighbor by letting the water flow from the neighbor's paddy into his during the night, and he would have gone back out and closed the opening before dawn if he had not been killed.

Stealing water was a very serious matter, if for no other reason because of the immense amount of labor involved in getting the water into the paddies. To move and raise the water from the level in the drainage ditches up and into the level of the paddies, the farmers used a tripod made of branches, a rope made of vines, and a piece of woven matting that was shaped like one half of a cone. They would swing the cone back and forth and when it came forward, it would catch a small amount of water, probably not more than a gallon, and "slop" it up onto the paddy. They would stand there moving the water for hours every day, for what seemed like several weeks. From our interactions, they appeared to pray for rain, but they only had the "cone method" for moving water.

One of the most memorable occasions of interactions with the Catholic Villages was when a local priest asked for me to come to his village and meet with him. I will admit I wondered if this was going to be legitimate, but I went to his village anyway. In what was one of my most meaningful memories of my life, I was taken into a small thatched church. It actually had a tapered roof and was open on the end opposite from the altar. In the middle was a small table where, with immense pride, he had placed a glass bottle of Coca Cola for us to drink together. That was the first time I had consumed a "hot" Coke. It must have been at least 100 degrees. I tried to show him my sincere appreciation for his generosity. He showed me how the dried mud was cracking along the walls, roof, and end of the church near the altar. He said artillery fire was damaging his church and he asked if I could do something to stop it. I assured him I would look into it, and I would do everything I could to assist him with the problem.

What appeared to be causing the damage were artillery rounds as they went over the church, rather than by artillery rounds actually hitting nearby. The church was in an area where artillery fire did not have to be used to control enemy activity. I determined there was an artillery battery in the proximity of the village, and when a mission was fired in a direction where the rounds would pass over the village, the high velocity of the artillery rounds would generate small "sonic boom" vibrations that would hit the church. The sonic vibrations were apparently causing the mud to crack. I asked Col. Love to possibly intervene, but realistically, nothing could be done, and it was agonizing for me.

The Vietnamese people who farmed the rice paddies were displaced from the areas outside a "magical line" beyond which everyone was considered to be an enemy combatant. This boundary line only extended a few miles inland from the South China Sea coastline. The indigenous people who lived outside of the line were displaced to "camps" where they lived in squalid conditions by our standards, while some might have considered it better than their native habitat.

I observed a tragedy one day while we were being transported to the 2/7 Headquarters in trucks. When we passed a massive trash dump created by the United States forces, it looked like an ant hill because of innumerable Vietnamese people digging in the trash in search of food or anything else they could salvage. A young child quickly moved forward in the pile and picked up what appeared to be an open C-ration can. It apparently had some uneaten food inside. An adult Vietnamese grabbed the can, and when the child resisted, the adult viciously struck the child in the head with a board. The child appeared to be instantly knocked unconscious, or was possibly killed. As we watched, the child laid there motionless without anyone paying any attention to it. We eventually lost sight of the horrible incident. Such was the reality of war.

In another horrific encounter, my troops notified me that a young child had been wounded during a firefight we had just engaged in. They said the child was lying by a water well a short distance away. When I got to the well, I found a small child approximately two or three years old lying on the ground with a bullet wound to its head. No one was there, but the child was still breathing. I asked Cathi to call a medivac chopper for the child. As I held the child in my arms, I wondered what I could have done to have prevented this senseless act of violence against this innocent little human being. I also wondered what had happened to the parents that caused the child to be left behind. As I held the baby while waiting for medical help, I felt its life slip away. The death of that precious little girl has haunted me my entire life. I had Cathi notify the medevac chopper that it was too late, and because of the responsibilities of my command and the ongoing contact with enemy forces, I had to lay her lifeless body by the water well with the hopes that someone would come for her. I can only imagine the heartbreak suffered by her parents or survivors of her family when they found her lying there dead...and alone.

Collateral damage is a horrible consequence of any armed conflict, and innocent people are usually the victims. As a partial explanation of the events that could create one of these senseless tragedies, it is necessary to understand how artillery and mortar rounds were fired in those days as a military practice. Artillery rounds and mortar rounds were propelled by the ignition of gun-powder increments, which were different sizes of supposedly waterproof plastic bags of the explosive material.

When a fire mission was called, the units firing the shells would determine the distance to the target, and then they would place the correct number of powder increments inside the tube, or on the shell or round, to make it travel that distance. The tube or barrel of the weapon would then be adjusted to the proper angle so the explosive would hit the target. The variables for artillery fire included wind, weather conditions and, as I described earlier, the elevation of the target. If the powder increments had somehow absorbed moisture, it could result in the explosive material not igniting properly, and the end result was that the round or shell may not go as far as it was supposed to in ideal conditions. We called that malfunction a "short round."

When you consider that mortar and artillery projectiles could be adjusted to travel less than 100 yards in the case of a 60 mm mortar, or to travel as far as 22 miles in the case of a 175 mm artillery round, complications could occur, especially when speed and time are of the essence. As you can imagine, one of the first things that Company Commanders were taught was to avoid calling for artillery or mortar fire that had to travel over their units on its way to the target, to diminish the possibility of a short round hitting our own troops. In the case of final defensive fires, it was not of any concern, because when you called for that fire mission, it meant you were being overrun. The final defensive fire mission also meant that the friendly forces were hopefully down in their holes and the enemy forces were above ground, so the odds would be that more enemy would be killed and wounded than our troops would be.

It was never more clear to me than when, on one particular day, I received information that an artillery round had apparently not traveled the intended distance and it had impacted in a village in my TAOR. The information also contained the allegation that there were civilian casualties. I took Cathi with me to investigate the situation. We discovered that it was apparently a short round that landed woefully short of its intended target, possibly by as much as a mile or two. Whatever the cause, it was virtually impossible to determine what went wrong or which artillery unit actually fired the round. After we arrived, I was taken to where the remains of two children were lying in a hooch. The children were so young that they had still been nursing sustenance from their mothers. Cathi described the scenario in his book as follows (page 392):

> In this case, I accompanied Captain Chamberlain as he went to this ville to ascertain the damage. What we found was that the round had landed on a hooch and killed two infants. The skipper talked with the village chief and gave them a few hundred piasters in compensation. I don't know how much he gave them but it didn't amount to much. And with the small payment, the problem was resolved as much as it was ever going to be. The reimbursement of a paltry sum for human lives did not escape us, but there wasn't much we could do. Most of the people whose children were killed never got anything.
>
> But the image that stayed with me was the sight of the mothers of the two infants. They were seated side-by-side on the ground, rocking back and forth, and wailing with profound grief. Each woman had her swollen breasts exposed and was methodically squeezing the milk out of her breasts to relieve the painful pressure. Though they were mothers of infants, neither looked young. Two grief-stricken, old mothers wailing and

squeezing their milk onto the ground-----this is the image that comes to mind when I think of the pain of noncombatants who live in a war zone.

To offer Vietnamese piasters as compensation for the lives of these two precious little children still makes me nauseous these many years later, especially when these villagers lived in a barter society where money had no relevance to most of the indigenous population. To make things worse, I could not even communicate our genuine sorrow to them in any meaningful way because neither of us spoke the language of the other. Despite all of the sorrow, however, I did hope that we had been able to diminish somehow the horrible agony of the two South Vietnamese families.

"Huey"

HUEY WAS A CLASSIC EXAMPLE of the positive, meaningful effects I believe we had on at least one young South Vietnamese. He was a fourteen-year-old boy who somehow had become connected to Echo Company before I became the CO. He was probably less than four feet tall and weighed an estimated 75 pounds. He was very fluent in English and he was a very valuable asset to us in numerous ways. He would appear whenever we were in the area around Hill 190. When he appeared, he would live with our Company and he was literally my eyes and ears as far as the Vietnamese people were concerned.

I took every measure I could to protect his identity and what he was really doing for us. He would interpret for me with village and hamlet leaders, and he would give us intelligence information he gathered by listening to people in the area. However, his most important activity was listening to the Kit Carson Scouts converse with the village and hamlet leaders when they wanted to speak to me. On numerus occasions he was able to tell me what was actually said and revealed that the Scouts were not always being truthful. It allowed me to have Scouts replaced when they had been untruthful, and on several occasions Huey saved us from being attacked or ambushed.

As you might imagine, he was a "hero" to us, and we talked at length about bringing him back to the world with us. Sid and I ordered and paid for a special sized pair of jungle boots for him...and he wore them for about thirty minutes. He had never worn anything but his "flip-flops," and he could not make the adjustment.

Occasionally, an older Vietnamese gentleman would come to our location and try to do anything he could for us to earn money from us.

He often told me he wanted to thank me for the C-rations or other things we were able to give him. I told him he did not have to do anything for me, but to prove his gratitude, he showed up one day with a Vietnamese woman who was dressed very provocatively, like a woman might be in an area of DaNang where "sex was for sale." I was surprised and concerned why someone like that was brought to our location. I had Huey find out what was going on. He told me that the "Popason," as we called him, had brought me the woman as a favor and that it was a way of thanking me for letting him work for us. She stared at me with what I thought was a look of hatred, and obviously I was shocked by what was happening. I was able to have Huey explain to both her and the Popason that I thought she was very beautiful, but that I was not interested.

Once again, Huey had "saved the day." Huey was very amused that I did not accept the "gift" of the woman from Popason, and with a grin, he kept saying to them and anyone else who would listen, "Skipper say no boom-boom! Skipper say no boom-boom!"

I have wondered about Huey's fate for these many intervening years, and I can only imagine that he had to be one of the first targets for the Communist North Vietnamese when they overran South Viet Nam because of his relationship with my Company. Wherever he is and whatever his fate, Huey was a great friend to us and appeared to be a tremendous patriot. He had the leadership talent that could have possibly revolutionized the world if he had been in a position to lead. May God bless you Huey, wherever you are.

"wbp"

When I was growing up in rural Wyoming, and as it is today, the power generated by internal combustion engines was/is described as "horsepower." I never have understood precisely where that particular nomenclature came from, but it appears that it is a comparison between the measurable power of a particular engine as compared to the power of a single horse or number of horses. Prior to the industrial revolution in the United States, horses in our country were generally classified as either saddle horses, which were ridden as a means of personal transportation and working with livestock, or draft horses that were used to perform the work necessary to accomplish agricultural production. The calculation of the power of a horse as it is compared to the torque of an engine is baffling to me, but it has achieved a commercialization value that I am sure will last longer than I do.

With that said, I decided to describe the measure of the power unit for agricultural production in South Viet Nam as a "wbp." The acronym I created means "water buffalo power." The historical significance of the water buffalo in the agrarian Vietnamese society was illustrated daily by the role these magnificent animals fulfilled in the daily lives of the indigenous rural populace in South Viet Nam.

Water buffaloes served as both the means of transportation and the power necessary to produce the rice that was the sustenance of the people in the countryside. The critter was fitted with a harness that basically was a "wood yoke" that allowed the powerful animal to pull things. To prepare the rice paddies for planting, the water buffaloes were used to pull a curved piece of wood, or stick, about six to eight feet long through the rice paddies. It would have the effect of a "plow" that would stir the soil to a depth of six to eight inches. The angle of the stick was controlled by a farmer walking behind and holding onto the end of it. During the next operation in the preparation of the seedbed, the farmer would stand on a log approximately three to four feet long with stubs of branches sticking down into the ground four to six inches. As it was pulled over the surface by the critter, the result of the operation left the surface of the paddy smooth enough so rice could be planted in it. The planting was done by hand. The amount of moisture in the soil determined how successful the operation was, so the timing of the farming operation had to be coordinated with the moisture precipitated by the weather patterns. If the weather did not cooperate, it could lead to attempts to steal water that I mentioned earlier. The water buffalo was also used to pull wood carts with two wooden wheels to haul things, or to be ridden on by people who were unable to walk, needed assistance, or children who decided to ride them.

I would estimate that the average water buffalo weighed between 900 and 1,000 pounds. I do not remember ever seeing a baby water buffalo, and I never knew where the trained, mature animals came from. I did not notice if the adult buffaloes were cows (females), or bulls (male), or neutered bulls (steers). I also did not see any indication of how ownership could be proven, such as the branding of animals like we do in the western United States and have done for many decades. It did not appear that everyone, or every family, owned a water buffalo, but it did appear that there was one or more around each hamlet or village.

The children were the ones who seemed to tend to these big animals. It was fascinating to see these large creatures so domesticated that the child tending them could be sleeping on their backs while they grazed in or around the rice paddies. All of the water buffaloes had a ring in their nose with a small rope attached to it, and the child in charge of the critter would often lead it around. On occasion, the water buffalo would swim out into eight to ten feet of water in one of the drainage canals or ditches, and the child would stand on the head of the buffalo between its horns. As it swam, all that could be seen above the water was the nose and horns of the buffalo and the child standing between the horns using the rope that was tied to the ring in the buffalo's nose for balance. It was an absolutely amazing thing to see. Unfortunately, we also saw the results of the tragedy that occurred when one of these large animals stepped on a mine or booby trap, and especially if a child were involved.

Along with the fascinating part that the water buffalo played in the lives of the rural Vietnamese people, we quickly learned that those same water buffaloes wanted nothing to do with us. For some reason, whenever we approached one of these creatures, they became irritated very quickly, and if they felt challenged in any way, they would charge us in a similar way bulls do the clowns and participants in a rodeo in the world.

Cathi described one of his encounters with a water buffalo in his writings about our operations during the Tet Offensive (page 460-461):

> The following day, we went into the village of Nam O and searched it house to house. My only moment of excitement that day occurred in the backyard of one house; the yard was basically a corral, and it had a water buffalo in it. The big animal was spooked, first by the firing and then by people coming into its quarters. I was following the skipper with my radio on, and I was carrying my .45 pistol in my hand since we were in close quarters. Suddenly, the water buffalo charged through the group of Marines and came straight at me. After seeing one take a hundred bullets the previous summer, I knew the futility of trying to stop a water buffalo by shooting it, so I didn't even try. Instead, I emptied my .45 into the dirt in front of me, and the water buffalo veered around me and ran out of the enclosure.
>
> Everyone laughed at my firing at the ground, but I knew it made more sense than firing at the buffalo and possibly annoying him.

Not only did Cathi's actions spare the life of the animal, he preserved one of the most essential things the Vietnamese people in the countryside needed to be able to survive.

Operation Pitt

As the movement of enemy troops into the DaNang area became more obvious, the mission of my Company was directed even more at intercepting enemy troops moving down Elephant Valley from Laos. This is part of what led to my Company being moved more frequently into the Hill 190 base camp and the division of my TAOR that went over the Hai Van Pass north to Phu Bai. The final decision was made to divide my TAOR making the northern boundary the actual Hai Van Pass and to move my Company to Hill 190 effective 28 December 1967. The northern portion of my TAOR was placed under the tactical control of a different unit. However, prior to implementing the change, Operation Pitt was launched on 5 December.

Col. Love had decided that we should return to the general area of the mountains to the west of Hai Van Pass where he had led the small operational group soon after he took command of 2/7 that I described previously. (See **Learning the Actual Tactics and Operations of a War** chapter.) The preliminary search of the area soon after I took command of Echo Company had convinced me that a trail we discovered was a significant route used by more than those who were authorized by the South Vietnamese. Our intelligence sources said that the trail we had identified went to the top of the mountain range and then curved back toward Hai Van Pass to the east. If that were the case, the mission of my Company in this particular Operation was to move westward up Elephant Valley and then attack northward forcing any enemy forces we encountered upward and eastward on the trail. Col. Love and the rest of the partial Battalion forces he had committed to the Operation were going to try to locate the east end of the trail and push westward from the Hai Van Pass area in a "pincer" movement until we had the possible enemy forces trapped between us. If we did not make contact with enemy forces, we would link up somewhere on the trail as we moved our units toward each other.

When my Company had initially discovered the trail, it was an eerie experience. The large size of the trail was surprising, and we had a foreboding sense of danger that made "the hair stand up on the back of your neck." During the short time we were on that trail initially, we had an eerie feeling that we were being watched, but we did not have the opportunity to engage any enemy forces at that time. As usual, the NVA/VC chose not to engage my Company on our terms at that time, which possibly was because they did not want to bring attention to the looming Tet Offensive and the part that trail might play in their future attack plans on the greater DaNang area of operations.

Wet weather covered the mountains on each side of Elephant Valley as we moved out of our base camp launching Operation Pitt, but it did not last long. Because of the dense jungles, re-supply became a problem. The matched parts problem with the M-16 rifle (See **The Military Industrial Complex** chapter.) became exemplified during our first re-supply. There was a triple layer of jungle canopy, at least 100-150 feet high, where the Operation Plan indicated we would receive the supplies in a "landing zone." Quite obviously, that was not going to happen. I decided to use some of the C-4 explosives that I always ordered my men to carry. We were able to make a hole in the canopy by cutting trees off near the ground with the C-4. We used a M-14 rifle to shoot a smoke grenade up near the opening to mark our position. The re-supply was to be done with two of the older CH-34 helicopters. Our position was on a sloping "finger" on the side of that particular mountain, and we had one platoon cleaning rifles down the slope of the finger away from the opening in the canopy. Cathi and I talked to the pilots on the radio and asked them to make their approach from either the east or the west across the finger, to hover, and then drop everything through the spot where the smoke could be seen, if they could see it. They confirmed they could see the smoke. We also told them we had troops down the slope of the finger and to make sure they did not drop anything down the slope from the opening.

The first chopper was able to hover, and the crew dropped their load by throwing everything through the hole in the canopy. It was difficult because of the limited power of the CH-34, some wind above the trees, and the tremendous heat further reduced the power of the choppers. As we found out later, the pilot of the second chopper had only been in-country a few days and this was his first mission. When he tried to hover near the opening, he was unable to do it. Instead of going around for another approach and coming across the finger as we had cautioned them to do, he turned partially around and approached coming up the slope of the finger. By the time he realized that he did not have enough power to fly up the slope of the finger, all of us were in trouble.

I heard the chopper approaching from the wrong direction and it sounded like it was attempting to fly up the slope. Suddenly, there was the sound of a burst of power from the engine. By this time, he was directly over part of my Company, including me. In a matter of seconds, I could hear the rotor speed starting to slow down and I could hear the chopper starting to come down through the trees. The foliage and brush were so thick, none of us could escape quickly in any direction. Chunks of trees, limbs, and rotor blades were showering down through the layers of canopy, which fortunately slowed the downward speed of the chopper and the many broken pieces of debris, providing us some protection. As the chopper crashed and finally hit the ground, the motor was still running slowly, with the rotor blades broken off. Cathi and I were desperately trying to dig into the dense foliage to get away from the crashing chopper but we could only move a foot or two before it hit the ground on top of my men. The rear of the chopper where the tail rotor is located was so close to us, we could reach out and touch it. In my opinion, it was only by the grace of God that the aircraft did not explode before the pilots turned off the engine. However, it landed on top of 14 of my Marines who were cleaning their M-16 rifles. None of them were killed, but all of them were seriously injured.

We now had 14 rifles that were field stripped with the hundreds of possible combinations of parts that would have to be compared and reassembled. This was during the time when the CH-46s were still grounded, and our only way to remove the injured was with the use of an Air Force HH-3E "Jolly Green Giant" helicopter and a jungle penetrator, which was not the primary mission of the Air Force. The highly skilled, courageous Air Force pilot hovered there for over 25 minutes lifting out my injured troops, the pilots, and the crew of the downed CH-34 involved in the crash. Fortunately, and miraculously, the pilots and crew of the downed chopper were not injured. I later learned that the Air Force chopper had an engine fire warning light on the whole time the pilots continued the lengthy extraction. I recommended the pilots and crew of the HH-3E for Distinguished Flying Crosses for their bravery and service to my Company, but I was told that the Marine Corps did not recommend personnel from other branches of the United States Military Service for decorations. Another chopper came later and lifted out the pile of M-16 parts and the radios that were in the chopper that crashed on us. We destroyed the remainder of the downed chopper with C-4 explosives before we moved up the large trail the next day.

Col. Love and his units could not find the trail we believed to be near Hai Van Pass. However, they did find one that was much smaller, and it did not appear to go to the area of the jungle we had entered. The result was that our forces were split, and the mission was becoming much more tenuous. If my Company encountered the size of force I thought was possible from the indications we were finding on the trail, combined with what I knew from our previous reconnaissance, I believed our mission was seriously compromised.

Col. Love decided that we should join forces as soon as possible. He told me the movement of his force was very slow because of the density of the jungle and the small trail they were using. He gave me a time to be where he believed our trail would connect with his, which would have necessitated that I move my Company quickly on the larger trail we were on. However, we were not sure the trail we were on would ever intersect the trail he was on.

After my experience when we were ambushed on a trail shorty after Col. Love took command of 2/7, I had decided to never move quickly on a trail unless it was a matter of extreme danger, or "life or death." I told him during a secure radio transmission that I believed it was in the best interests of my troops for me to attempt to navigate through the jungles by my use of a compass and keep my Company off the trail since our plans had changed. I do not think he thought I could do that, given the quality of the maps and the density of the jungle. Quite frankly, I confidentially had my doubts also, but I preferred to use my skills I thought I had in land navigation instead of possibly moving too quickly in an area where I had an overwhelming feeling that we could literally "get our heads handed to us."

My experience of growing up in rural Wyoming was where land forms are "home," and the people there move from one land form to another and describe locations based upon proximity to land forms. My additional experience with land topography as a private pilot also gave me the confidence to inspire approximately 180 Marines to follow me. My decision caused us to resort to the World War II method of traveling more safely in the jungles of the Pacific theatre... and that was with the use of machetes.

It was very difficult, laborious, and challenging, but one thing I was confident of was that we would not be ambushed. The relief map I had was supposed to indicate changes in elevation every time the elevation of the land changed 20 meters up or down. I only knew where Col. Love and his force thought they were, because he was using the same antiquated map I was. However, the map indications of the locations of the old logging roads that were built by the French during their occupation of this area decades earlier had proven to be fairly accurate in the past.

I was doing what was referred to as "dead reckoning," which meant I was trying to move my Company in a straight line as much as possible from where we were to where I thought Col. Love and the other force was, with only the use of a small, handheld compass. I stopped a couple of times and we had everyone remain motionless and silent while I would call for an artillery battery to fire an illumination round to a map coordinate where I thought we were. Even in the density of the jungle and canopy, we hoped to be able to hear the "pop" when the parachute and flare were blown out of the shell casing canister, which was sometimes possible. After we heard the first one that was at least in the proximity of where we were, my confidence began to build that we were not going to be lost forever in the jungles of South Viet Nam.

A few hours later, we found a logging road that I thought was the one I was looking for. It was nearly impossible to discern, except for the level surface about 15-20 feet wide. The estimated map locations that Col. Love and I used fortunately proved to be accurate and, after six to eight hours, my Company walked into Col. Love's defensive perimeter. Once again, the enemy forces chose not to engage us, and they avoided any contact. Operation Pitt was closed after our forces were returned to their base camps.

Months later, after 2/7 had become the Special Landing Force, Marine units that replaced my Company in the Hill 190 area read some of the intelligence documentation that we had compiled and left for the use of whichever units were operating in that area. They planned an Operation utilizing helicopters to drop the Marine forces on top of the mountain range instead of going up the large trail my Company had been on. According to an Officer I met who was a Company Commander on that Operation, there were very heavily fortified enemy positions at the top of the mountain designed to defend against forces like ours coming up the trail. He said an attack going up the trail would have been disastrous, especially because air support and artillery could not penetrate the heavy jungle canopy in that area. This was Divine Intervention again, in my opinion.

Christmas, 1967

AS THE TRADITIONAL CHRISTMAS DAY approached, it became very obvious to us how time, days of the week, and special days had lost much of the traditional significance and celebration back in the world. We had almost lost track of exact days and months, because it did not make any difference what day it was. Our efforts against the Viet Cong and North Vietnamese had to continue 24/7/365. However, the one thing every Marine knew was how many days he had before he rotated back home to the world, because we all had our "short-timers'" calendar.

The calendar was a picture of something, like a sports car, airplane, or voluptuous woman, to signify the number of days in a 13-month tour that Marines had left to serve in VN before going back to the world. The picture was divided into very small segments with numbers inside each segment. The numbers ranged from 1-395, and the idea was to fill in a box chronologically in reverse order each day to indicate how many days the Marine had left in-country. One thing no one ever was allowed to touch without permission was the short-timer calendar that belonged to someone else. With that said, many of us did not realize which day was actually Christmas Day.

We were on a search and destroy mission in a large rice paddy area when Cathi notified me that we were supposed to form a "360," or large circle, because we had a "V.I.P." coming in. We quickly formed a large circular area to provide as much security as we could, and a helicopter landed in the middle of my Company. The pilots shut off the engines and a Catholic Priest stepped out and told me he was there to offer communion to anyone who wanted to participate in the religious ceremony in celebration of the birth of Jesus Christ.

In addition to that surprise, the crew of the chopper got out and one of the pilots was my college friend from John Brown University, Captain Mike Bryant. He told me when he found out the mission was to take the Priest to my Company, he volunteered for it. As it was the first time when I saw him earlier in November, my spirit was renewed, and I was overwhelmed that he would do such a great thing for me and our friendship. We sat under a tree and visited for a while and then he had to move on to other units, but I will never forget how much I appreciated him spending that small amount of time with me, and how it gave me strength to carry on.

I never imagined that I would not see Mike again for another 35 years, and how our lives would change through time. Obviously, I did not realize the looming 1968 Tet Offensive was just over the horizon...a stage of the Viet Nam War that would change the outcome of the war dramatically, and change the history of that part of the world significantly.

Tet

IN AN ANTHOLOGY PREPARED by Brigadier General E. H. Simmons (Ret.), he quoted United States Army General William Westmoreland, the Commander of United States Forces in Viet Nam from 1964 - 1968, as saying 1968 was going to be "the year of decision." On **January 1, 1968**, of the entire United States and Allied Forces in-country, **81,248** were **United States Marines**. The harsh reality was that from **January through June of 1968, 3,057 Marines were KIA (Killed in Action) and 18,281 were WIA (Wounded in Action)**. From **July through December of that same year, an additional 1,561 Marines were KIA and 11,039 were WIA**. On average, **13 Marines were killed and 80 Marines were wounded every day in 1968**. In the first six months of the year, **17 Marines were killed and 101 were wounded every day**. Analyzed differently, **of the total number of Marines in-country at the beginning of 1968, nearly 6% were KIA and 36% were WIA, a combined combat casualty total of 41.8%, during the next 12 months**.

Major General Donn J. Robertson had been in command of the 1st Marine Division since 1 June 1967 and continued in that position into 1968. After the Tet Offensive, a significant change was made on both the political and military stages in the United States. President Johnson removed Robert McNamara as Secretary of Defense on 1 March 1968 and replaced him with Clark Clifford.

With that preview, the new calendar year of 1968 began with steadily increasing enemy activity. In January, leading up to Tet, many enemies killed in our ambushes were wearing civilian clothes instead of the traditional VC black "pajamas," and they were carrying larger than normal backpacks. Their packs included military uniforms, larger than normal quantities of rice, and unusually large supplies of ammunition. They were avoiding contact with us as much as possible and they appeared to be reinforced with Chinese units with personnel similar to the ones we had killed near Phu Bai. It was very obvious that these troops were fresh and unfamiliar with the terrain because of their physical size and appearance. The mistakes they made navigating through the countryside and jungles indicated they were not familiar with the area. Intelligence reports we received from 2/7 were becoming more ominous. We were regularly warned of increased enemy movements of large numbers of troops through the infiltration routes from the Ho Chi Minh Trail in Laos into the DaNang theatre of operations.

Complicating the "surge" of enemy troops was the other ridiculous decision that had been made by the Johnson Administration that I previously mentioned. The civilian Commander in Chief had mandated a halt to all bombing of the supply routes north of the Demilitarized Zone, or DMZ. The alleged intent was to "offer an olive branch" to the North Vietnamese and the Viet Cong in an attempt to get them to scale back the conflict. That tactical order had apparently been issued the month before Christmas, and the effects of the cessation of the bombing began to be felt in earnest by infantry units like mine in January 1968.

In retrospect, and in light of Secretary McNamara's book, I have often wondered if this was not an intentional act to allow the Tet Offensive to occur. It could have been a perverted attempt to create more support for the war effort in the United States. However, some critics have suggested it was done to allow our enemies to inflict so much damage on the United States forces that it would provide the Johnson Administration the opportunity to declare that we could not win a land war in Asia. If that were true, the rationale could then have been that the United States would have an excuse to begin a withdrawal from the conflict. After what we had seen, done, and not been allowed to do there, nothing would have, or would now, surprise me.

In January 1968, we also began to see the new 37 mm anti-aircraft guns mentioned previously that were being employed by the NVA and VC. They were very effective against American helicopters. As I stated before, the new transistorized FM communication radios that were apparently made in China were being used in the enemy operations against us. Many of the weapons were so new the cosmoline used to protect the metal during shipment and storage had not been completely cleaned off of them. Everything pointed to increased military activity and an escalation of the war. In reality, we were unknowingly in the final preparations of a coordinated attack against our American and Allied Forces throughout the entire country of South Viet Nam. It was leading up to possibly the largest military operation in the history of North and South Viet Nam.

Toward the end of January, my Company had been moved north of Phu Bai. The possibilities of the Tet Offensive had become more obvious. The afternoon before the Tet Offensive began, I had been given orders to build a blocking position south of Hue at the junction of Highway One and a smaller road going to the west. The urgency of the order was such that we were making a desperate attempt to fill sandbags and construct them into somewhat of a defensive perimeter as quickly as possible. Late in the afternoon, a group of CH-46 choppers appeared, landed, and offloaded an entire Company to replace us. With very short notice, my Company was ordered to load onto the same choppers, and we were flown directly back to Hill 190 to be repositioned in the Rocket Belt west of DaNang. The only explanation I was given was that we were moved back there because we had more experience in defending the Rocket Belt than other units. Several months later, we heard that the Company that replaced us south of Hue had been almost entirely killed and wounded during the launching of the Tet Offensive that began later that night. It was just another one of the many blessings that I felt protected my life during my tour of duty in South Viet Nam.

When we landed at Hill 190, it was nearly dark. Everything seemed to be basically the same as it was the last time we were there. The 81 mm mortar crew set up quickly in the sand bagged pit, and the rest of the Company moved into their holes around the defensive perimeter of the base camp. I had received several messages during the afternoon that were classified, and I could not share them with the troops. I tried to keep Sid apprised of what I was learning as much as I could, but that was as far as I could legally go with the status of the communications. The last message was particularly ominous, and well beyond the normal intelligence messages. It basically said that a massive attack from forces much larger than my Company was imminent, and that I could not rely on any support or reinforcement from artillery, aircraft, or other ground units. The bottom line was my Company was on its own, and our mission was to attempt to stop any enemy movement or activity involving the launching of rockets from the Belt.

I did not think it was tactically wise for me to send out patrols after dark since we had been dropped into the position without enough time for me to recon the area. I needed time to be able to make tactical decisions concerning the most effective way to deploy my troops. In addition, I could tell from the tone of the message that Col. Love knew our only chance to make a difference would be to attack enemy forces that may be near us after first light.

Information gathered by the VC and NVA, when they had overrun American military units in the past, usually included the names, addresses, and pictures of family members and friends of the family members of the U. S. forces who had been killed. The identification of people mentioned in the personal papers of the military personnel sometimes appeared in the news media back in the world, and that information was often read by Hanoi Hannah during her radio broadcasts which could be heard by United States forces in South Viet Nam.

To explain in more detail, the Hanoi Hannah propaganda broadcasts used by Jane Fonda, and traitors like her, were used to berate American Soldiers, Sailors, Marines, and Airmen. We were often referred to as "baby killers" and "murderers of the Vietnamese people." With that in mind, I passed the word for everyone to burn all their letters, pictures, and anything else that could identify their family and loved ones back home. I did not have to explain why. My order said it all, and my Marines in the Company instinctively knew we may not live through the night.

We tried to make sure that every field of fire was absolute, and extra mines were distributed throughout the perimeter where any ingress might likely be attempted. All of it was done under the cover of darkness. I know anyone who wanted to could have gotten close enough to us to hear the preparation, and I have always believed that what happened next was very likely because of the excellent work of my Marines. We waited, ready for anything and everything...and we waited, straining every ear and eye in the unit...and we waited some more.

After what seemed like an eternity, a rocket barrage began to hit the DaNang Air Base. However, the rockets were launched from areas other than in the part of the Rocket Belt we were trying to defend. Because our position was at an elevation high enough that we had line-of-sight vision of the DaNang area, we could see some of the immediate effects of the attack. Even though we were approximately eight miles inland from DaNang, the fires and explosions began to light up the night sky. The intensity of the huge balls of fire and flames continued to grow as daylight began to show in the eastern skies.

Our position had not taken one round of incoming, and I believe it was partially due to the excellent preparation of my Staff NCOs, NCOs and troops. However, as the rice paddies became visible, we could not believe our eyes. As I mentioned before, our position overlooked the Song Cu De River on the north and sat at the end of a long ridge sloping from the west to the east along the south side of the river. The rice paddies stretched to the east and south of our base camp, and the paddies had numerous small North Vietnamese flags along some of the paddy dikes that were used for foot traffic. They were small and attached to bamboo slivers, and the slivers were stuck in the ground along the trails. The NVA units had chosen to bypass our position and, apparently, the villages and hamlets in our TAOR.

The reason may never be known, but it appeared to me that either they did not want their movement to be discovered by engaging us, which would have alerted U. S. forces at the DaNang Air Base that an attack was under way, or they thought the engagement might be too costly, given my unit's excellent defensive preparations. I also believe it was, again, Divine Intervention. Whatever my belief or the enemy's actual reason, it was obvious that we were now trapped behind enemy lines with nothing more than the resources we had, and help was not, and would not be, on the way.

The fires at what appeared to be the DaNang Air Base were overwhelmingly visible, even in the clear light of day. A couple of hours after sunrise, I received a request for a condition report from 2/7, and when we answered that our condition was secure, it almost sounded like the person on the radio was surprised. I was directed to keep our position consolidated until I received further orders. It seemed like we were in the back row in a theatre, and all the "bad guys" were sitting in the eight miles of seats in front of us. The first day was uneventful, and I could only imagine what was going through the minds of those Vietnamese in the villages and hamlets around us. They, too, had been spared, but I was very concerned about what had been required of them to keep them quiet the night before.

Fires and explosions raged in and around DaNang during the second night. The light from the fires was so bright that it illuminated our position almost as bright as day, only with an orange and red color. We had no enemy contact the second night and began the second day with orders to maintain the defensive perimeter. I was becoming very concerned with the limited communication from 2/7 and our instructions to maintain our perimeter. It appeared to me that we were in a position that 2/7 could not help us or they could not figure out a plan for us to move and consolidate forces with another unit.

During the third night, the fire at the DaNang Air Base was still raging, but we could not hear or see evidence of any more explosions. Early on the morning of the third day, I received a terse message from Col. Love that simply said, "It is everybody for themselves. Try to get to the ocean to get picked up." Recalling again that DaNang was about eight miles east of our position, I notified my PCs and NCOs to get the troops ready to move. The closest shoreline was at the village of Namo approximately three miles east of our position.

Shortly after that, I received another message that I was to prepare part of my Company to move into a blocking position west of the village of Namo located north of DaNang on Highway One. The message said that units from the South Vietnamese Rangers would be attacking the NVA and Viet Cong who had overrun the village in the initial Tet Offensive assault. I had discussions with Sid, and we decided that I would take two platoons with me to carry out the blocking position mission, and I would leave Sid in command of the remaining platoon and mortar units. Their mission would be to defend the Hill 190 base camp.

Our blocking position was going to be far beyond the range of our mortars. In addition, the last 2/7 message directing me to assist the South Vietnamese Rangers included notification that we would not be able to receive any artillery support during our execution of that mission. Sid and I had agreed that once I completed the blocking position and had taken whatever actions I was ordered to do following that, we would decide how we were going to proceed. However, the original message I had received about our needing to get to the ocean for evacuation did not seem possible considering the requirements of the blocking position mission.

My operational plan was to move part of my unit eastward down the Song Cu De River with the use of the two amtracs I had at Hill 190. The rest of the unit would move across the rice paddies in a sweep formation. I wanted to utilize whatever camouflage we could to conceal our movement from the enemy forces located in the coastal areas while we were moving into our blocking position. Cathi described the operation in his writings as follows (page 453-455):

> We rode in amtraks down the river toward the village and got out of the tracks about a click and a half (1500 meters) from the village. Then we hiked out into the middle of the dried

rice paddies and set up our ambush. We laid behind a long rice paddy dike, with the Skipper and me in the middle, and a platoon extending out on each side of us. The place where we set up seemed arbitrary to me but I was soon to be reminded that Captain Chamberlain knew what he was doing. It was the most bizarre ambush I ever saw. Here we were, broad daylight, 60 Marines in helmets and full gear, lying out in the middle of a huge open field with no cover other than the paddy dikes.

Fighting erupted in the ville. We could see smoke and hear the firing. Planes started flying close overhead--the old propeller driven Skyraiders flown by the South Vietnamese Air Force. And then, amazingly, a large group of men came walking directly toward us. As they got closer, I could plainly see their weapons--they were holding their rifles down by their sides to hide the outline from the planes. A large group of these VC walked directly into the center of our ambush. It was hard to believe that they didn't see any of us. I could see all these Marines lying prone with their rifles aimed directly at these guys and not one of them saw us. I guess they were too preoccupied with the planes flying overhead.

The skipper and I were dead center in the middle of the line, and the platoon commanders, along with their radiomen, were located at each end of the line. The Skipper said, "Tell everyone to fire on my command."

I passed the word over the radio. Everyone exercised good fire discipline and held their fire. Then when the group of VC was about twenty-five yards in front of us, the skipper gave the command to fire, and I relayed it over the radio to the platoon commanders. Our two platoons opened fire and shot down eighteen guys in one volley. None of the VC even got a shot off. Neither the skipper nor I fired our weapons at that point because it was obvious we didn't need to.

After that, the VC knew we were out there, and we no longer

had the surprise factor. Indeed, it was now our turn to be the targets. We had to make our way across the huge expanse of open field toward where the main body of VC was positioned, and they were shooting at us every step of the way. We got on line and began moving across the field. The VC were dug in behind a line of paddy dikes that was on the other side of the field, but they were still separated from the ville by a large flooded ditch that ran behind the ville. We advanced across the field by sprinting from dike to dike. The skipper would wait until everyone was lined up behind a dike, then he would give the signal and we would all get up and charge to the next one, where we would flop down behind the next dike.

When we first started across the field, we could hear the bullets around us but no one was getting hit. We were a long way from their positions, probably eight or nine hundred yards or more. But after we got about halfway across, Marines started going down every time we sprinted toward another dike. We weren't hitting any of them yet because they didn't have to get up and expose themselves the way we did. And we couldn't even stop to tend to our wounded; we kept going and left it to the corpsmen to take care of them.

At the end of the afternoon, we had a body count of sixty VC. We had quite a few casualties, though I don't believe any of ours were killed.

The fight at Nam O was the most killing I saw during my entire tour. More people may have died at other times, but I had not witnessed it directly as I did that day at Namo. I saw almost every one of those VC as he was hit and went down. By that point in my tour, I could find no energy to expend on feeling bad about the VC who died. It is only now, many years later, that I mourn everyone who died that day. It's no longer about us versus them; it's just death, lots of violent death.

After the South Vietnamese Ranger units had withdrawn, we moved into position to make a "cleanup" attack on the village. A Marine who was a Hell's Angel back in the world made a "banzai" frontal assault on two machinegun positions and single-handedly destroyed them, killing all the enemy fighters in the two positions. He had suddenly become enraged after his buddy had been wounded.

It was then that we began to notice Marine helicopters flying around in different directions. We had casualties that we had been carrying with us, and we called on the aircraft frequency for a medevac. The answer we received totally bewildered me. The response to my request was, "Negative on your request. We have VIPs on board." We were then able to see television cameras sticking out the doors of the Marine choppers with the letters CBS, NBC, and ABC on the sides of the respective cameras. The American media had been given precedence over wounded and dying Marines!

We finally took control of the village of Namo, captured numerous weapons, and reached the ocean, which was DaNang Harbor. When I notified 2/7 of our location, I was told to maintain my position and await further orders. I talked to Sid on the radio and updated him. I told him I would let him know whenever I knew anything more about him moving the rest of my Company to the ocean where we were. After approximately an hour, I received orders to move my Company back to Hill 190. A vehicle came to Namo Village to pick up our wounded Marines, and we completed the move back to our base camp just before dark. On the way back, one of the amtracs ran out of gas and we had to leave it to possibly be recovered later.

While the decision for us to return to Hill 190 seemed difficult to understand, it is important to remember that the size of the Tet Offensive was enormous, encompassing all of South Viet Nam. The resulting property damage was so immense, and the killed and wounded on both sides so numerous, those at the higher levels of command had to issue orders quickly. Then, as the fluid situations often changed rapidly, many of those orders had to be changed, and some changed again.

Our return to Hill 190 consolidated my Company, but it was difficult to be an effective Marine Rifle Company offensively. We were still isolated, and we were not being re-supplied. 2/7 made it very clear to me that we had to get along with what we had for the foreseeable future. I asked the Gunny and Sid to inventory the ammo bunker. Cathi has the following recollection of the meeting and wrote (page 465-467):

> *It took a while for the seriousness of the situation to sink in. One afternoon, Captain Chamberlain let me sit in on a meeting with the platoon commanders and the platoon sergeants. He started by announcing that he had asked the gunny and the XO to inventory the ammo in the ammo bunker. He held up a sheet of paper.*
>
> *This is what we're down to.*
>
> *One of the platoon commanders took the list and whistled. The XO said, 'That's right. We're almost completely out of mortar ammo and we're close to the end of the M16 and M60 ammo.'*
>
> *Captain Chamberlain said, "I've decided to distribute everything that's left."*
>
> *The Gunny said, "All of it?"*
>
> *The Captain said, "Everything. Make sure every bunker on the perimeter gets its share of the Claymores and pop flares. That stuff will just stay in the bunkers. We'll divide up all the M16 ammo and the grenades among the rifle platoons. The gun teams will take all the machinegun ammo, of course, and they've already been assigned to permanent bunkers. Do the same with everyone on the hill. I want every man on this base--the cooks, the drivers, everyone--to know exactly where he goes when we get hit."*
>
> *One of the lieutenants asked, "Don't we want to have some central supply depot that people can go to? What if we get hit on one side of the perimeter and use up our ammo on that side?"*
>
> *Captain Chamberlain said, "If that happens, we'll be shifting personnel from the other sides, and they'll have to carry their*

ammo with them. No, there's no sense leaving anything in the ammo bunker. From here on out, everyone has got to make what they have last, because that's all they're going to get."

There are always things that do not get taught in Marine Officer Basic School. What does get taught is to make the best decision you can based upon the principles you have been taught in training. But even more importantly, make decisions based upon the particular situation you are facing by utilizing your best judgement at the time the decision has to be made. I tried to do that, and fortunately it was successful for my troops and their families and loved ones back in the world, which always weighed heavily on my mind. We made sure, again, that everyone had destroyed their personal information, and we maintained a primarily defensive position while we waited. Supplies finally came several days later, but they were delivered by a truck, since helicopters were in very short supply.

It was some time before we learned the extent of the damage to the DaNang Air Base. In actuality, there was a period of time when it appeared that the enemy forces would overrun the Air Base, but they failed. Reports were that several hundred dead NVA and VC were in piles along the fences surrounding the huge American aircraft facility. Some of them were probably enemy forces that chose not to engage our position on their way to a much more lucrative target. We began our routine patrolling in the Rocket Belt, but the "die had been cast."

Even though this failed Tet Offensive by the NVA and VC resulted in a huge loss of life to their forces, the anti-war sympathizers in the United States gained the necessary publicity to bring political pressure to bear on the United States government. The primary intention of the anti-war movement was to force the eventual withdrawal of military troops from South Viet Nam. That did not happen until over 58,267 young American lives had been lost, hundreds and hundreds of thousands of young American's bodies had been wounded, and innumerable minds of American youth had been shattered by the atrocities of war.

For the first time, I realized that our beloved Country was headed in a direction that someday would result in its self-destruction, unless drastic measures were taken...and yes, I felt a loss of patriotic fervor. The horrible treatment of Viet Nam Veterans in the years following the War was deplorable. It was accentuated, in my opinion, because of how expendable our lives as Soldiers, Sailors, Marines, and Airmen were to the pathetic, traitorous leadership of our Country at the time.

Incidentally, as for the Hell's Angel who had acted so heroically on our way to the coast, he was supposedly a convicted felon who had been given the option by a sentencing judge: either go to prison or join the United States Marine Corps. He had a hand nearly deformed from years of karate and breaking concrete blocks with that hand, and he did not fear anyone or anything. When there was enemy contact, "he was at his finest," and his actions near the village of Namo were evidence of his bravery and courage.

As for the body counts that were so important in the Viet Nam War, I do not know for sure how many enemies were killed near Namo Village that day. On my Silver Star nomination form that I first saw nearly 50 years later, it mentions a number of enemy KIAs while personal accounts of some of my men say considerably more. Whatever it was, I do understand the necessity of accounting for it. However, it has always seemed "foreign" to me that the killing of numbers of human beings is a reason that I should be honored...

Leech Valley

AFTER THE TET OFFENSIVE, enemy activity decreased dramatically. In a tactical move, American forces became more aggressive after inflicting heavy losses on the NVA and VC during the failed Tet Offensive. We were sent on a Company search and destroy operation to an area we called "Leech Valley." It was west of Hill 190 along the Song Cu De River part way up Elephant Valley toward Laos. As you might imagine, the name "Leech Valley" was a prophetic term for what we would experience when we went there. This area of river bottom land was covered with dense elephant grass which was a narrow, single-leaf plant that grew straight up from the ground to a height of at least 10-12 feet. These long leaf structures were three to four inches wide, and the edges were sharp enough to cause small lesions on your skin, especially your hands, as you pushed through it. The ground under it was always very moist or had shallow standing water in it.

    The namesake of the area was created by the leeches that covered the plants. Most leeches would get on your clothing and move to a secure area like your crotch or armpit. In Leech Valley, there were numerous leeches on every plant, and when you pushed through the dense forage, the "blood-sucking varmints" would immediately penetrate your skin whenever they touched it. As a result, they would immediately bury their head in your face, your arms, and your hands. After they had filled with blood, they would fall off and the wound where they were attached would bleed. The Corpsmen said the bleeding was our body's way of cleansing the wound, which in turn reduced some of the related infections.

    I will always remember my first encounter with a leech in the first months I was in Viet Nam. It was in an area other than Leech Valley, but on that day, I looked down and my entire crotch was soaked with blood. Needless to say, I immediately "dropped my drawers" and saw the spot where a leech had been. The spot was bleeding at an alarming rate I thought...and the blood was coming from the side of my left testicle. Since we did not wear skivvies or T-shirts, the blood came through the lightweight material of my jungle utilities very quickly.

Pulling a leech off of your skin was dangerous because when you did that, the head of the leech normally stayed attached to your skin and it could easily cause infection. The thing that worked best to remove the "critters" was some very powerful liquid insect repellent that we used for protection from mosquitoes. A small amount of the repellent would make the leech pull its head loose and it could be removed. Too much repellent would kill it and the problems would magnify, not to mention the horrendous pain in the "private areas" of the body if that skin was burned by the repellent. Some troops tried to have condoms sent from the world to wear when we were in areas like Leech Valley. However, most said their request was usually met with skepticism and suspicion, especially since most of their families and/or loved ones had never "been there, done that." As you can imagine, a leech lodging itself in the head of a penis could prevent the individual from being able to urinate, and that was an absolute emergency medevac. I actually did see that happen to one of my Troops.

During one of our movements through Leech Valley, I had my Company in three lines about 25 yards apart. I was in the middle line, and we were stepping over a log lying in the mud and grass. Someone finally stepped on the log...and it moved. The Marine was so spooked he cut loose with a burst of gunfire as he dove into the mud, and everyone else dove into the mud, too. When we finally established that there were not any "bad guys" around, we began to check out the "log." It turned out to be a giant python, or some other kind of huge snake. It was 12 steps long and had a large hump in it. We tromped the elephant grass down all around it. Being totally aware that stopping for very long could endanger us, I still let the troops tease the enormous monster. When we killed the snake and cut it open, the hump inside the giant serpent was a calf that could have weighed as much as 150 pounds.

I still have nightmares about that snake. I did not relate that story to anyone until the early 1990s. I was invited to Taiwan as a representative of the State of Wyoming, and I had heard many stories about "snake alley" in Taipei. While there, it was confirmed to me that some Taiwanese actually raise snakes of that size for use in religious ceremonies. I felt vindicated because until then, I was not sure if the snake in Leech Valley was as large as I remembered or if my mind was playing tricks on me.

Snakes were deadly in the jungles. The bamboo viper was small, bright green, and so deadly that we did not carry any anti-venom for it, if it was even available. We were warned that if one bit your hand and you had a machete in the other hand, the only way you might survive would be if you cut off your bitten hand instantaneously. The cause of death from the bite of a bamboo viper was from the onset of paralysis of the entire nervous system of the victim, often within a few seconds after being bitten, according to the warning. When the respiration ceases due to the paralysis, the victim actually dies of suffocation. Some people referred to the bamboo viper as the "two-step" snake, meaning you could probably only take two steps before you began to die after being bitten.

We moved on through Leech Valley because our primary mission was to try to find elephants that may have been used to transport military supplies, especially artillery pieces, not to marvel at the size of snakes. We finally encountered a small group of elephants, though not enough to call a herd. As we tried to get close enough to see if there were pack marks on the elephants, one of them began to charge in our direction. My troops began shooting every kind of weapon at this huge beast. Suddenly, the elephant fell to the ground, so some of my men carefully approached him. He suddenly stood upright, turned, and ran away from us. Needless to say, we were relieved but, unfortunately, we could not be sure if we saw any pack marks.

Ice Cream?

DURING ONE OF OUR SUPPLY DROPS in the spring of 1968, one of the helicopter crew said he was supposed to ask us if we wanted some ice cream. He went on to say that supposedly the Marine Corps was planning to build a facility that could be used to make the delicious product. How much more like the world could that be? I was skeptical, but I told him we would take some if it ever became available. I did not think any more about it, and then one day a few weeks later, I received a message that a chopper would be inbound to our position in about 30 minutes with some ice cream. Since we were moving on a Company-sized search and destroy mission, we moved into a defensible position and waited. It was a blazing hot day, as usual, and I stopped the Company out of curiosity as much as actually expecting ice cream to be delivered.

As the chopper approached a few minutes later, I wondered how the "ice cream people" knew where we were and, of course, how we were going to be able to eat it. Ice cream bars? As the CH-46 landed, the pilots shut the engines down so the dirt and dust would settle. Unbelievably, in the back near the ramp was a 10-15 gallon metal "vat" full of ice cream! No lid, melting fast, but VANILLA ICE CREAM! The people who sent it must have thought we were at a base camp or somewhere where we had utensils...but that was not the case! Time was of the essence, for several reasons. Leadership at every level, in my opinion, requires that the Troops eat first. Officers and Staff NCOs stood aside while the Troops walked by...and grabbed a handful! As might be imagined, the melting of the ice cream and the dirt and grime from the many hands going into the vat produced a pool of mud and "slop." The sight of it might have caused a discerning individual at the end of the line to give pause and decline the invitation "to have some ice cream." I can assure you such was the case, because I was at the end of the line.

The recognition of those who were involved in this effort to do something special for us was not possible because, regretfully, we had no idea who they were. As the chopper flew away, we were very thankful for the thoughts of those behind the delivery. We moved on with our mission, despite the Troop's sticky hands, and we were grateful for the brief reflection on the life we yearned for back in the world.

Non-Combat Casualties

DURING ONE OF THE SEARCH and destroy missions in the mountains northwest of Da Nang, we encountered a small black bear. He was not aggressive, but we quickly shot him, "gutted" him, and carried him to a nearby village. We gave him to the villagers, who we thought were Montagnards, or "Yards." You would have thought we had given them a semi-truck full of juicy steaks. Because they did not have conventional weapons, they probably would never have been able to take the bear. Many Yards had supported U. S. forces during the war, and I think we won some more "hearts and minds" that day.

Shortly after giving the bear to the villagers, one of my Radio Operators stumbled into a hornet's nest. Instantly, these huge hornets attacked him and three other troops who were close to him. The Radio Operator did not have a shirt on, and the hornets swarmed down between his back and the radio he was carrying. We could not get close to them, so I threw a smoke grenade in on top of the troops who were on the ground writhing in pain and trying to fight off the huge insects. The smoke worked and the hornets rose above the colored cloud, swarming in the air above the smoke. We were able to get into the smoke and drag the troops away to a safe distance. By that time, the Radio Operator was in convulsions as we tried to get the radio off of him. He lost consciousness, and I feared for his life. As we waited for a medevac, we counted over 100 bites on his back. The emergency treatment by the Corpsman almost certainly saved his life, and two of the other three recovered rather quickly. Once again, the helicopter did not have anywhere to land, but the expert pilots helped save the life of the Radio Operator and a second Marine with the use of a jungle penetrator. As I said before, someone trying to kill you was only one of the difficulties that had to be dealt with to survive in South Viet Nam.

On one search and destroy mission, I had my company dig in for a night defensive position surrounding a hamlet. The Vietnamese who lived there had suffered immensely from the torture and killings carried out by the NVA and VC prior to us moving into the area. After our holes were dug, some of the troops noticed a small puppy in the hamlet and started playing with it. It was chewing on their fingers and hands as puppies do. A couple of hours after they began playing with it, the pup began to foam at the mouth, and it started having convulsions. It died within an hour or so, and we immediately suspected rabies. After checking with medical personnel back at 2/7 Headquarters, we believed the only safe thing to do was to medevac the seven troops who had physical contact with the cute little pup so they could receive rabies shots for their protection.

Unsung Heroes

IN MY OPINION, two groups of "unsung heroes" in the Marine Corps infantry units are, as I have alluded to before, Radio Operators and Corpsmen. As most people know, the United States Marine Corps is an organization within the United States Navy, and Corpsmen are members of the Navy that are attached to the Marine Corps. The Marines are the "ground troops" of the Navy. While the Navy Seals now often conduct operations on land, they are a dedicated, professional unit separate from the Marine Corps. The Navy Seals have "come into their own" in the last 30 years or so, but they were not that apparent during the Viet Nam Era, as far as I knew then. The Marine Force Recon units were who we thought performed the clandestine operations in the on-land warfare arena during the Viet Nam War. They were only vaguely similar to what the Seals are today.

We could not have had more professional, courageous, and self-sacrificing heroes than those who served as Radio Operators and Corpsmen. As I described earlier, both the Corpsmen and Radio Operators were targeted by enemy forces who tried to "take them out" first. Anything that would interfere with our communications or interrupt the medical treatment of our troops was of vital importance to the North Vietnamese and Viet Cong. The belief that medical personnel were not targeted by enemy combatants is a "fairytale played out in the movies." Marking a Corpsman with a white cross to supposedly keep them from being shot by enemy forces would have been "issuing him his death warrant" in South Viet Nam.

As with Radio Operators, we did everything we could to make Corpsmen look like everyone else for their protection. I had both groups carry M-16 rifles. However, the last thing I wanted Radio Operators and Corpsmen to do was to have to use their weapons, because that most likely would have interfered with their primary responsibility. Radio Operators carried their radios in their packs with the microphone cords strung down through their sleeves, and the flex antennas were shoved down between their packs and their backs. We made every effort to conceal their identity.

Radio Operators and Corpsmen constantly had to expose themselves to enemy fire, and their performance had an immense impact on the outcome of every contact with enemy forces. In addition, in the case of the Corpsmen, they were responsible for the day-to-day physical readiness of every Marine in our Company, including me.

I remember one day when I began to feel weak, had a headache, and gradually became nauseous. At first, I thought it was just the extreme heat, but before long I "came to" after having passed out, and I realized one of our Corpsmen was trying to get me to swallow water. He told me briefly what had happened with my labored breathing and loss of consciousness. He thought it was a lack of salt in my system, which was a dangerous problem, considering the quality and quantity of water we consumed every day. While there was an enormous amount of salt used as a preservative in the 25-year-old World War II era C-rations we ate, I apparently needed more. The Corpsmen helped me recover by putting salt in my water canteen regularly. These young medical personnel had only the basics of emergency medical care with an emphasis on the treatment of wounds to the body, but they also had an immense sense of logic and "moxie" in their approach to the medical care they provided.

Radio Operators and Corpsmen will always have my deepest admiration and respect, and they were and are special heroes to me. None of us realized how much their training, courage, and resourcefulness would be tested in the not too distant future. My Company would be one of two companies to initiate one of the longest lasting Marine Corps Operations in the Viet Nam War.

Lives Changed Forever ...Operation Worth

THE GENERAL TACTICS used by Marine ground units did not seem to be more aggressive in the spring of 1968 after the Tet Offensive. Search and destroy missions were carried out regularly, but contact was sporadic in most areas. However, enemy contact southwest of DaNang appeared to increase steadily. As mentioned before, the area west of Chu Lai and south of DaNang was supposed to be under the operational control of the South Koreans, or ROKs. As in the U.S., anti-war demonstrators were steadily increasing pressure on the South Korean Government to withdraw their military forces from South Viet Nam. The political unrest in South Korea seemed to create apprehension on the part of the ROKs to actively carry out normal military operations. It also appeared that the unintended result of the ROK's reluctance was creating a supposed "safe haven" for the VC and NVA in the area from Chu Lai west to An Hoa. Part of the large safe haven area was Go Noi Island near the "Happy Valley" region.

Intelligence sources reported that the huge area had become a virtual R&R and re-supply area for the NVA and VC because of its proximity to the Laotian border and the Ho Chi Minh Trail. The unintended result of the ROK reluctance to act was that the adequately supplied and rested NVA and VC troops were initiating new offensive operations. These enemy operations were unexpectedly coming from that area southwest of DaNang, including the Rocket Belt, and surprisingly just a few months following their defeat in the Tet Offensive.

A very significant change had just taken place in 2/7. Col. Love had been re-assigned and was now the Executive Officer, or XO, of the

7th Marine Regiment. I had been flown back to the 2/7 Headquarters to honor his service in a small, colorful, farewell ceremony to give him his final salute. It was performed to the music of the Colonel's favorite song, "Sammy Small." The serenade was provided by a small four-piece musical group of Marine musicians. Without further elaboration or explanation, the lyrics to that little musical "ditty" will not be reduced to writing in this book. Suffice it to say, the lyrics may seem appropriate if Marine participants have been partaking of alcoholic beverages, but they would probably defy the imagination of most reasonable people who are not Marines. In our tradition, the music seemed very appropriate for loyal 2/7 Marines to honor this great Marine leader during the change of command ceremony...without the lyrics and alcoholic beverages.

Col. Love was replaced by a "Mustang" who was in the latter stages of his rise through the ranks after beginning his career in the Marine Corps as an enlisted Marine. The new 2/7 Battalion Commander was Lieutenant Colonel Charles Mueller, and he had every reason to be proud of his personal achievements. However, there was a dramatic difference in the manner and style of command between the two officers, which was understandable.

I was the senior Company Commander in 2/7, having served in the capacity of Echo Company CO under Col. Love for five months. I do not know if the new Battalion Commander was suspicious of the loyalty of officers in my position or not, but it did appear to me that he was somewhat paranoid concerning my length of command experience in South Viet Nam, and my rapid promotion to the rank of Captain. If my memory serves me correctly, I do not believe this new Battalion Commander had ever led a Marine Rifle Company in combat.

Soon after Col. Mueller took command of 2/7 in March 1968, my Company was deployed to the area just north of the ROK's TAOR on a search and destroy mission. Our deployment into this area on Operation Allen Brook would happen later in May of that year. On this particular Operation, which I later learned was named Operation Worth, we had the usual orders to "search, engage, and destroy the enemy forces, and their supplies and facilities." Be that as it may, our deployment on this particular mission was as a Company operating alone under the command and control of 2/7, but in coordinated efforts with other units. As night approached near the end of the second day of the Operation, I moved my Company onto a ridge position overlooking parts of Happy Valley. We dug in where we had excellent defensive visual observation capabilities and fields of fire. In a situation like we were in, we normally were given the radio frequencies of other units in the area for tactical, logistical, operational, and emergency uses, if it became necessary.

As a result, one of my Radio Operators was monitoring the frequency of a Marine Battalion located on a ridge west of our location after darkness had set in. My Radio Operator told me he thought I should listen to an open conversation, meaning not coded or "scrambled," between the Battalion Commander of First Battalion, Seventh Marines, which was located west of our position, and the Seventh Marine Regimental Commander. The 1/7 Commander was describing to the 7th Marine Regimental CO an ongoing engagement between his unit and a "superior enemy force." The information that the 1/7 CO was relating to the Regimental CO was that they were "outnumbered," but that they were "going to fight to the last man." In the open grassy area where we were located, we could see the ridge where 1/7 was located. Even though we were at a distance of one to two miles from the 1/7 position, we could not hear any gunfire or explosions, and we could not see any tracer rounds, which I thought we probably should have been able to see in the darkness, and possibly hear from our proximity to their location. At the time, quite frankly, I wondered if it was an exaggerated report.

Soon after the radio transmission ended, I was contacted personally on one of my radios by the 7th Marines Regimental Commander, as had been the case when we had taken "friendly fire" because of a battery zone fire gone awry several months earlier. The Regimental CO described the desperate position 1/7 was in at their location west of us, mirroring what he had been told in the conversation I had monitored. He told me I was to go to their defense at first light the next morning. I did not comment other than to acknowledge the receipt of his order.

Little did I know that the experience of the next few days would change my life and the lives of many of my troops...forever. In my mind, the foundation of Marine Corps history, pride, and tradition was smeared and stained beyond repair during those next few, short days. My sense of guilt and failure still haunts me. From this point on, I am leaving it to you, the reader, to draw your own conclusions. In addition, I hope the disclosure of the following information will lead the United States Marine Corps to reveal the truth about what happened during the chain of events that I will describe to you. The fact that Headquarters Marine Corps in Washington, D.C. had for years officially claimed that my Officer Qualification Records, or OQR, were missing had reinforced the appearance of a "cover-up," in my opinion. I hoped that was not the case, and as I wrote this, my intense desire was that the records concerning the following tragedy would somehow be released in their entirety.

The choppers that would airlift us to the landing zone or LZ where we could begin our efforts to reinforce 1/7 were supposed to be at our location by first light. However, I received a message that the "pickup" would be delayed because they were "prepping" the LZ. It became readily apparent that we were being dropped into an area no one had reconnoitered, particularly me as Company Commander. To prep the LZ, they were using "daisy cutters," which were 1,000-pound bombs retrofitted with a piece of metal several feet long that was attached to the nose of the bomb. As the bombs hit the ground, the metal rod would strike the ground first causing the bomb to explode just above the surface shredding the foliage around the blast. If it was successful, the ordinance would destroy enough of the brush so a chopper could land safely. After about a 30-minute delay, the choppers arrived, and I took my Radio Operators on the first chopper going in. The total "drop" of my Company was completed in about 45 minutes. Our landings were unopposed, and when I stepped off the chopper, I could readily see the challenge that was before us. The plan designed by those in command was to move us from our position on the ridge where we had been and drop us in the bottom of the valley between our previous position and the location of the besieged 1/7 Battalion on the next ridge. It was a typically hot day with estimated temperatures well above 100 degrees Fahrenheit.

The climb appeared to be over a mile in length and 1,000-1,500 feet in elevation. Considering the underbrush and partial jungle covered terrain, I was sure it would be a tremendous physical challenge, not to mention the unknown possibility of being attacked by forces above us as we climbed. My troops did everything possible to reach the top of the ridge before nightfall, and we did. However, it involved dealing with several weather-related casualties resulting in some troops having to carry the gear for those who had been overcome by the effects of the blazing heat. Those casualties had to be literally dragged to the top of the ridge. Our mission was to get to the top of the ridge and hook up with the beleaguered 1/7...and we did, but it was not pretty. Thankfully, we did not receive any fire from the enemy forces that had nearly overrun the Marine Battalion on the ridge.

At that point, I was no longer under the operational control of Second Battalion, Seventh Marines, and instead operational control of my Company was transferred to First Battalion, Seventh Marines. When I finally was able to inform the 1/7 Commander, or BC, that we were completely inside their lines, darkness was quickly enveloping the jungle. I asked if we could have an hour to rest up before we began whatever assignment he had for us. The BC declined emphatically. He pointed to a flat rock formation that was barely discernible in the impending darkness and said that was where they had the last enemy contact. He told me to deploy my Company there in a blocking position. I moved my unit to the location which appeared to be about 200 yards outside the main 1/7 position in a northwesterly direction. By the time we reached the rocky area, complete darkness had enveloped us. What we found was solid rock, and it was impossible to find any kind of defensive positions or dig holes.

I decided to deploy my Company in a "360," or circle, around the edge of the flat rock location, which was extremely tenuous at best. Because of the darkness, I thought I should have my PCs put out three listening posts around our position if we could possibly find a defensible position for them. Jack came to me a few minutes later and told me he had a Corporal that refused to take his squad out to a listening post position. The Corporal was one of the best, if not the best, Corporal in the Company at that time. I told Jack to have him come over to where I was. When he arrived, I told him to sit down, and I asked him why he was refusing to follow a lawful order. He confided in me that he had noticed that many casualties in our Company were happening in the 13th month of a Marine's tour. He also pointed out that he was in his 13th month, and he said he had a foreboding feeling that he would not survive the night if he went out to a listening post.

I had to make a quick decision, and one that was meaningful. I told him he had been an outstanding Marine since he had been under my command and that I would not try to force him to follow the order. In my mind, the last thing we needed right then was to have a confrontation that could have jeopardized the safety of my entire Company and possibly have created a distraction for the 1/7 Battalion to which we were now attached. The Corporal was black, but that fact did not enter into my decision.

Many people did not realize that Marines were in South Viet Nam for a 13-month tour. Some other US military branch's tours of military service in VN were less than that. As a result of the Corporal's actions, I contacted 2/7 Headquarters a few days later and asked that a survey of Battalion records be conducted to possibly substantiate his contention. The results of the survey were shocking. The survey indicated that 80% of the casualties in 2/7 occurred the first two months or the last two months of a Marine's tour in Viet Nam. As a result of the findings, a policy decision was made that was implemented for Marines in the rifle companies in 2/7 from that time forward. From the date of the policy implementation, those Marines in 2/7 rifle companies were sent back to the Battalion Headquarters to complete the last month of their 13-month tour. However, future statistical information did not help me solve the problem I was dealing with that night.

While I could have had the Corporal prosecuted for disobeying a lawful order, I am grateful to him for taking a stand on behalf of all Marines, and I have never regretted my decision concerning him. He had been instrumental in an incident that occurred when we entered a new base camp a few months earlier. In that base camp, there was a Confederate Flag nailed to some railroad ties next to my hole. He was with several of my black Marines who came forward soon after our arrival and demanded that I remove the Confederate Flag from the railroad ties. When I asked them why, they explained that to them it was the symbol of racism, slavery, prejudice, and every negative, unjust thing that their "people" had been subjected to throughout the history of the United States. In an abbreviation of the conversation, I asked them if they thought I was prejudiced toward them or had ever acted in a racist manner. They assured me that I had not as far as they knew. I told them it was not my flag and I had not placed it there. I also told them that if they did not like the flag there, they should take it down. After we visited about the various issues, they decided to leave it there. This Corporal had been a leader in that group, and he contributed immeasurably to my efforts to successfully resolve their complaint.

The night passed very slowly as we lay on the flat rock formation, but it passed without incident. None of the three listening posts had any enemy contact. At first light, I had patrols sent out to "recon" the area so we could make our best use of the terrain to establish a defensive position to support 1/7. It was not very long until one of the patrols made a "grizzly" discovery. I received a grim radio transmission from 2nd Platoon Commander, 2nd Lt. Jim "Pip" Pippen, who had sent out one of the patrols. Pip told me that he needed me to come to a bomb crater that was not far from the rock formation where we had just spent the night. Pip, his 2nd Squad Leader Corporal Joe Freda and 1st Squad Leader Corporal Joe Wnukowski were standing near something on the ground in the bottom of the bomb crater. As I approached them, I could see the shock on their faces. Those next few seconds of time changed my life forever. At their feet lay a partially decomposed Marine on a poncho. The Marine had been found by Cpl. Freda and one of his Fire Team Leaders. He was lying face-up, and his most significant injury seemed to be that most of his left leg was missing. What appeared to be his name was printed on his flak jacket with the use of a magic marker, which was a common practice in most units. I could see that his "dog tags" were still on a chain around his neck. I checked to see if the name on his flak jacket matched the name on the dogtags, and it did.

It appeared that the deceased Marine was being carried on the poncho by someone, or several people, when he had been left there for some reason. I could not tell if he was dead or alive when he was abandoned. In addition, there were open C-ration cans around the Marine, and the contents indicated that some of the rations had been only partially eaten. I felt shocked and frozen in place for a moment, not believing what I was seeing. Lying before me was the incomprehensible betrayal of a fellow Marine...an unthinkable rejection of revered Marine tradition, and damage of cataclysmic proportion to the very foundation of the Marine Corps. I could not possibly imagine how this could have happened. I told Pip to have his patrol continue on their missions and that I would deal with the situation.

As soon as the patrol left, I called the 1/7 command on the radio and told them I had a "Kilo India Alpha," or "KIA," meaning a Marine who had been killed in action. Coded language was always used for killed and wounded. I gave 1/7 the name, rank, and service number of this young Marine who had given his life in the service for his Country. At the end of my transmission, I asked for a medevac chopper.

The standard Marine Corps policy and procedure would have been for the chopper to take the remains to the morgue in DaNang for shipment home to the family or families of the dead Marine. To maintain full objectivity, and to be fair to the 1/7 Battalion that I had just become attached to, medevacs for Marine KIAs were normally not given the same priority as those that were requested for Marines Wounded in Action, or WIAs. I was holding the handset for the radio so the Radio Operator could hear the transmission, because I was immediately struck with the astronomical importance of what was happening. In addition, I wanted him to be able to carry on routine communications about the issue if I was doing something else. I also had an enormous foreboding of what had happened to this dead Marine, and I wanted to have a witness to whatever was going to take place.

I was unaware of who it was in 1/7 I was actually talking to on the radio, but such is communication by radio in most instances, especially when I had been attached to this Battalion for less than 24 hours. There was a lengthy pause, and then the voice on the other end of the transmission said the request for the medevac was denied. I replied that it was going to be impossible for my unit to carry the KIA with us and I said again that I needed, and was requesting, a medevac for the KIA. The voice came back and said, "Negative on the medevac, just bury him." That radio transmission has flashed back through my memory thousands of times since then.

My Radio Operator and I stood in stunned silence looking at each other in disbelief. I finally asked my Radio Operator, "Did you hear what he just said?" He almost unintelligibly said, "Yeah." I was speechless. Almost talking to myself I said, "What did he say?" With his lips trembling, my Radio Operator said, "He told you to bury him, Skipper." I stood there paralyzed with my mind exploding, and in a few more seconds the voice came back on the radio and asked me, "Did you receive my last?" I finally replied, "Affirmative, but I don't believe it." The voice then came back and said, "Do as you are told. Don't rock the boat. This is an order. Bury him." As I mentioned earlier, Cathi had already rotated back to the world. While not meaning to disparage the radio operator who had replaced him, my mind immediately "flashed back" to Cathi and how much I needed his assistance right then.

That was the most unbelievable, devastating thing that had ever happened to me personally. With the tradition, integrity, and honor of all the Marines who had preceded me and were serving with me, I could not have imagined that I would ever be faced with a situation of leaving a Marine behind, being told to bury him as part of a "cover-up," and to do it in the jungles of South Viet Nam.

I immediately tried to contact Col. Mueller at 2/7, but because we were so many miles from the 2/7 Headquarters and because of our location in the mountainous region, I could not reach him. I then contacted Sid and my acting Company Gunny Staff Sergeant Lutu. I told them what a mentally torturous decision I had to make. I also told Jack what was happening and received his input. I thought carrying the deteriorating remains with us would endanger my troops and possibly prevent us from being able to continue and complete our mission. After agonizing for a few minutes, I decided that the only possible choice I had was to bury the Marine in the bomb crater where we found him. I believed I could identify the bomb crater from the air. However, for the first time, I noticed that there were several bomb craters close by. As I walked the area, I thought I could identify the particular bomb crater I was going to use for the burial site when I returned...and I was going to return.

I decided that I would not allow anyone other than an Officer or a Staff Noncommissioned Officer to be involved in the actual burial. I asked both Sid and Staff Sergeant Lutu to assist me, since they had the rank criteria I had rather quickly decided should be involved. They both agreed.

Prior to the actual burial, I tried to talk to all the Marines in my Company to tell then what was happening and why I did not have the authority to do anything other than what I had been ordered to do.

I assured them numerous times that I would return and recover our fellow Marine as soon as we were back under the operational control of 2/7. The looks of horror, disbelief, broken trust, and betrayal in the eyes of my Marines that day have been seared in my memory forever.

Sid, Staff Sergeant Lutu, and I laid the Marine to rest. When I removed the first shovel of dirt from the potential burial site, I was overwhelmed by my inadequacy and inability to prevent this horrific event from happening. As we laid his remains in the shallow grave and began to cover him, my sense of failure and shame was the most powerful, negative, psychological trauma I have ever known. I was now complicit in an attempt to provide the cover-up of the truth about a Marine who had been left behind in South Viet Nam.

I believed then, and I believe now, that I failed everyone who has ever served in the United States military, and that I brought shame to bear on every Marine who has ever chosen to serve our beloved Corps. I have never been able to forgive myself for my actions that day. I chose not to leave a marker at the grave site because I was afraid it would attract attention from enemy forces that might then use it as a propaganda tool.

The trauma associated with leaving someone behind who is serving their Country is a direct violation of the policy of every branch of the military service, especially the United States Marine Corps. The abiding hope and trust that those who risk their lives for America have is that their service to their Country will be honored in life and death, and that the ultimate honor they will be given is the return of their remains to their families and their Country**...no matter what**. Recruitment tools of all branches of the military service recite the principle that "**No one will be left behind.**"

Unbelievably, I was ordered to violate the trust of everyone who has ever been a member of the military service of the United States and who has ever served our Country. In overwhelming mental anguish, I could not think of any alternative that I had except that I would somehow return to reclaim this Marine's honored remains. However, my stated resolve to do so did not compare to the devastation, destroyed confidence, loss of respect, and feeling of helplessness felt by most or all of my men.

We were released from the operational control of 1/7 the following day, and we were able to begin moving to a pickup point so we could be flown back to 2/7 Headquarters. As we moved past the bomb crater, I saw that some of the troops had made a cross out of small branches and stuck it in the ground over the shallow burial site. The tiny cross was adorned with a helmet liner and a set of prayer beads. At our Echo 2/7 reunion in May 2019, Cpl. Freda gave me the picture he had taken of Cpl. Wnukowski placing the cross he made on the burial site...two Marine heroes paying their final respects to a fellow Marine that they feared would be forsaken. May God forever bless them for their loyalty and honor.

My sense of guilt and failure as a result of this act so many years ago still makes it very difficult for me to sleep, to relax, or to be able to carry on a normal life. Being alone seems to be what I need much of the time, but that is also when my thoughts of self destruction plague me the most. I am sure there are, or will be, critics who think that I am making too much out of this "ugly situation." However, being despised by many Americans for serving our Country in Viet Nam caused us to cling to one concrete, time-honored promise and principle: we had been assured and believed without reservation that if we made the ultimate sacrifice for our Country, our remains would be sent home to our loved ones...and now even that was an apparent lie. At a reunion of our Company a few years ago, one of my men told me that my Radio Operator on that fateful day is confined to a mental institution in California, and has been there for many years...the price he is paying for my failure to somehow avoid the burial of our fellow Marine.

It took us all day to move to our pickup point, and in retrospect, I think we were distracted by the burial and probably were moving too quickly. Just after the chopper I was on lifted off, I saw a commotion among some of my men, but it did not appear that I needed to land, which would have been very difficult operationally, and I knew Sid was still in command on the ground. After we were back at the 2/7 Headquarters that night, Sid told me that one of our Marines had picked up what he thought was a hand grenade someone else had dropped, but it was a booby trap. He threw it quickly in an attempt to avoid injury to anyone. Sid said there were not any indications of injury to anyone, but tragically, our Marine immediately lost consciousness and was medevacked on the next chopper. I would see him at Tripler Army Hospital in Hawaii later in the spring when I went there on R&R.

The Rest of the Story

When the chopper I was on landed at 2/7 Headquarters, it was after dark. I went immediately to Col. Mueller's hooch to report what I had been ordered to do the day before. I knocked on the screen door, but there was no response. I inquired of a junior officer where the 2/7 Battalion Commander was, and he said, with a smirk, that he, meaning Col. Mueller, was in his hooch with his Vietnamese "girlfriend"...and he also said he would not be in any shape to talk to me until the next morning.

Needless to say, I was absolutely incredulous that an allegation like that would have been made or was possibly true. To have any Vietnamese civilian in the 2/7 Commander's hooch with all the maps, charts, and other military information would have been totally irresponsible, in my opinion. However, it brought to my mind recent discussions between Sid and me. We had remarked to each other that it sometimes appeared the NVA and VC knew in advance what we were going to do, but that is a subject for someone else to pursue. To be clear, I had no direct evidence that the allegation about Col. Mueller was true.

After spending a sleepless night, I was at his hooch again at first light. Col. Mueller appeared to be very tired, but that was none of my concern at the moment. He seemed a little irritated at my early arrival, and he poured himself some coffee while I began my accounting of what happened during the last three days. It seemed like "a light went on" when he realized I had been ordered to bury a Marine. He immediately notified the 7th Marine Regiment Headquarters.

As I mentioned earlier, the immediate past Commander of 2/7, Col. Love, was the new Regimental Executive Officer. Col. Love arrived by chopper and was standing in front of me in less than 30 minutes. In his typical style, he said, "What in the &!@#&! happened, Doug?" He did not allow Col. Mueller to attend our discussion, which I thought was odd. Col. Love's assignment as Executive Officer of the 7th Marine Regiment was undoubtedly going to be the deciding factor in my believability.

When I related the chain of events to him and told him I had 14 volunteers, including me, ready to go back and get the Marine, he did not hesitate. He told me as he left to go to the 1/7 Battalion Headquarters that the 1st Marine Division Air Officer would be there to see me within the hour. He suggested that I tell him what air support I thought I needed to expedite the recovery of the remains of the Marine I had been ordered to bury.

Approximately 45 minutes later, a Lieutenant Colonel from 1st Marine Division Headquarters landed in a helicopter, introduced himself, and asked me what kind of air support I needed to assist me in the recovery operation. I told him I was requesting two CH-46 choppers to carry seven Marines each and two Huey Gunships to provide cover if we were opposed.

A short time later two CH-46s landed while two Huey Gunships circled overhead. I had Sid stay in command of the Company and I took Staff Sergeant Lutu with me on my chopper, along with five other Marines from my Company, one of whom was Cpl. Wnukowski. Pip was in charge of the other six Marine volunteers from Echo Company that were with him in the second chopper, one of whom was Jack.

An indescribably grim atmosphere surrounded us as we stepped onto the choppers. Not a word was spoken on my chopper, but the eye contact between Marines was hollow and daunting. In less than 20 minutes we reached the area where the bomb craters were, and I could see what I thought was the burial site from the air. I had the pilot of my chopper notify the second CH-46 to drop his Marines near the bomb crater after we landed, and then to go airborne until we finished the recovery. My chopper landed on the edge of the bomb crater where the remains were buried, and the troops with me quickly scrambled off the chopper and took up defensive positions. Staff Sergeant Lutu and I jumped off the ramp on the back of the CH-46 and ran down to the burial site. Cpl. Wnukowski helped us dig up the remains of the Marine, and we carried him back to the chopper. My troops climbed aboard, and we lifted off. As the other CH-46 landed, Pip's Marines jumped aboard and the chopper lifted off without incident.

Once we were airborne, I asked the pilot of my chopper to thank the pilots and crews of the other three choppers for their assistance. I also asked that the gunships be released whenever our pilot felt they were no longer needed and requested that the Marines on the chopper that did not have the remains be transported directly back to 2/7 Headquarters. As it was when we started, the Marines on my chopper were silent, except that I could sense a fragile look of relief in the eyes of some of them.

When we landed at the DaNang Air Base Morgue, a team of Marines and Navy personnel carefully took the remains from us. We flew the relatively short distance back to 2/7 Headquarters in complete silence, except for the "whacking" of the rotor blades on the chopper. As my Marines exited the aircraft, I thanked each of them for volunteering to be on the recovery team, but none of them acknowledged me, except for Staff Sergeant Lutu. It had taken us approximately one hour after we started the recovery to return with our mission accomplished.

It took months for some of the details to emerge concerning the entire, sordid tragedy and the following attempted cover-up. I was sure I would never know the entire story. Even after 50 years, I can still remember the entire event that totally emasculated me patriotically, intellectually, and honorably, as if it were yesterday. As it was related to me by various individuals in 2/7, pilots, and from Col. Love in Hawaii several months later, the following is what was supposed to have happened when the dishonored Marine was left behind, and his remains were abandoned.

Supposedly, a few weeks prior to our discovery of the Marine, 1st Battalion, 7th Marines had conducted a Battalion-sized Operation in or near the area where we found the Marine's remains. 1/7 apparently had been in the area for a couple of weeks, and they had been engaged with enemy forces several times. They had been withdrawing after breaking contact with the enemy forces one day, and when they consolidated their position, a Company Commander reported to the 1/7 Commander that one of that particular Company's Marines was missing. It was also reported that some Marines in the Company said they had seen him step on an explosive, possibly a land mine, and that he had been injured. The 1/7 Commander then supposedly told the Company Commander of the unit from which the Marine was missing to go back and retrieve the missing Marine, because "We don't leave Marines behind."

In a "heroic effort," the Company Commander allegedly asked for volunteers to go find the missing Marine. According to the rumor, six Marines, the oldest of whom was 19 and all who were supposedly friends of the missing young Marine, apparently volunteered to conduct the "rescue/recovery" operation. The 1/7 CO arranged for them to have a radio to be able to communicate with a FAC for two F-4 fighter bombers that were providing "fixed-wing cover" and two Huey Gunships that were providing additional close-air support for the Marine volunteers.

As the story was told to me later, the rescue/recovery team, with the 19-year-old Marine in command, located the same Marine we found later, and also found his severed leg. The team of young Marines supposedly reported to their Company Commander that they strapped the portion of the leg on one Marine's back and placed the Marine on the poncho that they were going to use to carry him. No one ever told me whether he was dead or alive at that point in time.

During the return to the 1/7 location with the Marine, the rescue/recovery team supposedly stopped to eat in the bomb crater where we found the remains, which explained the open C-rations we found around the dead Marine. While they were eating, they were supposedly taken under fire by enemy forces. They then abandoned the Marine in the bomb crater and contacted the FAC. In the radio transmission they ostensibly told the FAC that they were under fire from enemy forces. They supposedly sent word to the F-4 pilots via the FAC that the enemy forces firing at them were around the bomb crater, and that the pilots should "bomb the bomb crater," which was the one where they had abandoned the Marine. The pilots or the FAC apparently could see the bomb crater from the air. That could explain the numerous bomb craters in the vicinity of the bomb crater where we found the Marine, but it also must have meant that none of the bombs hit the bomb crater where the Marine had been abandoned. It was apparently the plan of the six volunteers to have the abandoned Marine's body destroyed by the bombing as part of a cover-up that was to follow.

The Marines on the rescue/recovery team then supposedly returned to their Company at 1/7 with the partial leg of the missing Marine. The Company Commander was supposedly aware of what they had done, but he allegedly reported to the 1/7 CO that the leg was all of the remains of the Marine they could find and recover. At the time of our discovery of the remains, the family of the missing Marine had already been notified that he was Killed In Action, or KIA, and the partial leg had been sent home to his family. Allegedly, the family was told that the small piece of remains was the only part of their loved one's remains that could be found. I was also told that a funeral had been held by the family of the fallen Marine.

According to the information I was given from the various sources, the six Marines on the rescue/recovery team had all been recommended to receive bronze stars for bravery, and the two F-4 pilots had been recommended to receive Distinguished Flying Crosses. Supposedly, the Company Commander of the missing Marine had been reassigned to the 1/7 Battalion Headquarters sometime after the "heroic mission," with the assigned duties of being the S-3 Operations Officer, or the S-3A, for the Battalion. The reassignment of that Company Commander had taken place just days before my Company had become attached to 1/7 after we were deployed to reinforce them.

As the plot thickened, the new Operations Officer for 1/7 supposedly was the voice on the radio that ordered me to bury the Marine. Apparently, he chose to cover up the lies told by the six members of his

Marine recovery team that he had dispatched from his Company at the time the Marine had been left behind. It also appeared that because the former Company Commander knew what his six-man recovery team had actually done, he had possibly lied to his 1/7 CO, and that may have given him additional reasons to try to cover up the actions of his Marines. Up until we found the dead Marine, he also may not have told his Battalion Commander that he had not gone out on the recovery mission, choosing instead to send out the six young Marines. However, the conjecture can go on and on, but all I have wanted to know for 50 years were the facts and the truth.

Further information given to me indicated that once the Commanding General of the 1st Marine Division had the information about what had actually happened, the six bronze stars for the recovery team were canceled immediately so they would not be reported to the media as having been earned retrieving a missing Marine. It was never confirmed to me that the six Marines were disciplined for their part in leaving the Marine behind and participating in the attempted cover-up. However, it was rumored later that they were court martialed for their actions.

A reliable source told me later that when Col. Love went to the 1/7 Headquarters to investigate the incident after he had spoken with me, the Officers involved immediately began to lie, and they denied that they had any knowledge of me burying a Marine. They claimed that I was being untruthful, at which time Col. Love became so incensed that he allegedly punched a Marine Captain through the side of a canvas covered hooch while telling them he had me under his command for six months and he knew I would not lie to him. The various accounts also included claims that three Marine Captains in 1/7 were relieved of their duties and sent back to stateside duty as a result of the tragic, dishonorable betrayal and resulting cover-up.

A few weeks later, I was leading my Company on a search and destroy mission when I received a message, supposedly from the Commanding Officer of the 1st Marine Division, Major General Donn Robertson. The message said that he appreciated all I had done to accomplish the recovery of the Marine, and that he would like to recommend Navy Commendation Medals or Letters of Commendation for me and the other members of my unit who retrieved the Marine. The message also asked what my opinion was as to whether the Commanding General should do so. I thought then, and I believe now, that it was an attempt to win favor with us so that the issue would simply go away, and to complete the cover-up. Additionally, my understanding was that Navy Commendation Medals were not reported to the press. I declined, because I believed what we had done was what every Marine should have done under the circumstances, and our actions did not deserve any special recognition. I also believe that if the Commanding General had felt that strongly about the issue, he would have recommended us for recognition instead of asking me, the leader of the actual recovery team, for my opinion.

I have often wondered if the Commanding General of the 1st Marine Division even knew what had happened, and if the message had possibly been sent by an "underling" in a further attempt to conceal this atrocity. Little did I know that in a few short weeks and months later, more information concerning the issue would be related to me. By happenstance, it occurred when I was confronted by a superior officer who claimed that a United States Senator supposedly was now involved in the burial issue.

The agony I felt concerning the burial issue was consuming me on a daily basis. In retrospect, it seemed that initially I was experiencing shock, but as time wore on, my personal sense of failure became overwhelming. My command responsibilities helped keep me occupied, but I continuously tried to stay busy with something. Little did I know that the burial issue was going to be a "plague" that I would have to deal with for the rest of my life, and in what manner would I deal with it?

I mentioned earlier that I had been a high school and junior college basketball official for many years. While I was attending graduate school in pursuit of a doctoral degree at the University of Northern Colorado in Greeley, Colorado, I joined a local basketball officials association so I could officiate high school basketball in Colorado. Officiating duties were assigned by the organization, and I was assigned to officiate a game one night with Ralph Bozella, who happened to be the State Commander of the Colorado American Legion.

After making his acquaintance, he asked me if I would be interested in judging the Colorado American Legion Speech Contest. I said I would, and I served as a judge in that capacity for several years. One year at the Speech Contest, Ralph said he was going to speak to a local high school class near Loveland, Colorado about the Viet Nam War, and he asked me to join him. He said Robert "Bob" Allsop, another Viet Nam Veteran who had served as a Navy Corpsman with Marine Corps infantry and recon units, would be with us. I agreed to, but when it was my turn and I began to speak, I suddenly lost emotional control when I tried to relate issues about my service in South Viet Nam to the students. I had to discontinue speaking and sit down to recover.

Commander Bozella never asked me to participate again, but he made some phone calls to the Wyoming Veterans Administration, or WYVA, and suggested that they contact me for possible treatment. The WYVA did, and I had interviews with an American Legion Advocate and evaluations and treatment with personnel at the Cheyenne VA facility.

Operation Ballard Valley

AT THE CONCLUSION OF Operation Worth, and the travesty concerning the burial of the Marine, Col. Mueller notified me that we were going on a Battalion Operation that was again in the area southwest of DaNang near An Hoa. It was more of a "classical" infantry Operation with tanks and armored vehicles. My mind felt "fried" every day, but the constant mental requirements it took to run my Company, unbelievably, allowed me to keep my sanity. I had a continual need to always "keep busy until I was ready to drop," and that need has haunted me continuously since that time. Then, as now, by constantly keeping occupied, it has kept me from dwelling on feelings of insecurity, worthlessness, failure, and self destruction.

On 29 April 1968, 2/7 launched Operation Ballard Valley. At the end of the first day, our Battalion established a night defensive position that included positioning the tanks we had with us in a defilade position. This means the tanks were maneuvered into an old canal that had been dug during the French Occupation of Indochina. The only parts of the tanks that could be seen above the ground were the turrets. This allowed the 90-millimeter guns and the .30 caliber machineguns on the turret to be approximately three to five feet above the ground. They were aimed at an open rice paddy area that was approximately 200-300 yards wide, with a tree line along the opposite edge of the paddies. Needless to say, it was a formidable defensive position.

Col. Mueller briefed us as Company Commanders on the particulars of the defensive position. Part of the Operation Order was that no Company was to have listening posts in front of any other Company's positions. In addition, he ordered that the tanks located within each

Company's position were under the command and control of that respective Company Commander. The tanks had infrared sensing devices that could detect metal with accuracy for a range of at least 200 yards, depending on the terrain. The only exception to the distance across the rice paddies was a thin line of trees that extended toward the Battalion position to a range of approximately 150 yards. That extended line of trees was directly in front of my Company position. I chose not to deploy a listening post there because I believed that the tanks would have overwhelming fields of fire if enemy activity originated in that thin line of trees. In addition, I was being more cautious because I was not comfortable with the Company Commander next to me due to his lack of experience. He had only been in-country for a short amount of time.

A couple of hours after darkness engulfed us, one of my Tank Commanders contacted me by radio and said they had infrared sightings in the tree line in front of our position and that the sightings indicated movement toward us. I told him to "stand by." I called the Company Commander beside me on another radio and told him what my tanks were seeing in front of us. I told him I wanted to make sure he did not have any listening posts in front of my position. He assured me he did not, so I contacted the Tank Commander for an update. He said there definitely were several metal objects moving in the trees in front of us. I told him to commence firing. In an eruption of 90 mm and .30 caliber fire, the tree line was instantly shredded, but suddenly a red flare was fired into the air from the tree line.

The listening post that I had been assured was not in front of my position by the adjacent Company Commander was, in fact, there. Somehow, one of the Marines from his Company that was in the listening post had been able to fire the warning flare just as the entire listening post was being killed or wounded. The tanks instantly ceased firing. The adjacent Company Commander came running toward my position screaming that I had "killed his men." Quite obviously, he had made a horrible mistake...and I made the mistake of believing that his earlier statements about no listening posts in front of my Company position were accurate.

This was yet another tragic example of the death and wounding of Marines due to friendly fire. When I went to the Tripler Army Hospital in Hawaii while on R&R later in the spring, I tried to see if there were any survivors of this senseless slaughter. In addition, I looked for my Marines who had been wounded near Hai Van Pass in the ambush tragedy that had happened shortly after I took command of Echo Company. I will relate what I found at Tripler in a later chapter.

The organizational planning for Operation Ballard Valley was so tactically sound, and the armament and weaponry so powerful, that the enemy forces chose to only initiate contact sporadically. One of those times was when the amtrac "pigs" we had with the Battalion started to run low on fuel. Col. Mueller decided to have four of them go to the Marine facility at An Hoa to refuel and pick up some supplies to avoid the fuel and supplies having to be brought in by helicopter. The pigs were to be accompanied by two tanks.

The refueling plan appeared to me to be flawed from the beginning because of the enemy tactics we had experienced since the Operation began. I became very concerned when Col. Mueller told me he was putting Sid in charge of the mission. Sid and I had a quick conversation, and we immediately agreed that he would take one of my Company radios with him so he could communicate directly with me. Both of us had concerns that the 2/7 Battalion would not be able to react quickly enough if he ran into trouble, and we both believed he would. In addition, my Company was positioned on the side of the Battalion that would be closest to Sid and his unit if a reactionary force was needed.

Escorted by the two tanks, Sid only made it less than a mile, and his convoy of tanks and amtracs was ambushed. The NVA and VC were just waiting for us to give them the opportunity to attack a small unit. Sid had been provided very few ground troops for his mission, and the few he had were riding on the amtracs because of the speed of the vehicles. He called for help from our 2/7 Battalion, and then called me on our

Company radio. I notified 2/7 that my Company was in a position to assist Sid and that we could reinforce his unit immediately. The 2/7 Command approved my rescue plan and we moved out as quickly as we could in the brushy terrain. Sid had done a great job of "circling the wagons" and was able to defend his men and equipment until we arrived. We were able to reach Sid and his vehicles without sustaining any casualties, and the enemy attackers withdrew. My Company radio he had with him was invaluable for us to be able to coordinate the consolidation of our forces. Time was of the absolute essence. After we consolidated with Sid and his unit, I was able to suggest to Col. Mueller that I did not think that Sid's refueling mission could be completed. He allowed Sid to return his unit to the 2/7 position, and fuel and supplies were brought in by helicopter.

As 2/7 advanced on the planned direction of operation, our movement paralleled a river. The first evening, a sniper had "dinged" at us from the other side of the river while we were digging in. Whichever gook was shooting at us was a terrible shot, but the second night I told the 2/7 Artillery Officer that I would like to surprise our "sniper friend" that night. I pointed out a small clump of trees to him that was on the other side of the river. I speculated that if the sniper was going to try to use his marksmanship on us that night, I would guess it might be from there.

With that in mind, I asked our Artillery, or "Arty," Officer to register an artillery-fire concentration on those trees while he was registering our final defensive fires for our position. Final defensive fires are when a unit commander calls in artillery fire on their own position if they are being overrun by enemy forces. It was a tactic, when infantry units were forced to use it, that always seemed to improve their survival chances, and with less casualties. I believed in doing that every night, if possible. The tactical object of the final defensive fires is that the friendly forces are hopefully in their holes as the final defensive rounds detonate upon impact. Theoretically, the explosives should inflict the most casualties on the enemy forces who supposedly would be above ground.

Everything happened the way I had suspected with the nightly target practice by the sniper, and I asked that the concentration registered for the trees on the opposite side of the river be fired. However, when our Arty Officer called for that concentration, it only took a very short period of time for us to realize that the incoming rounds were not going to hit the target we intended. Instead, a barrage of 8-inch artillery rounds began to hit my Company position. The incoming rounds that were supposed to take out the sniper were instead our final defensive fires.

Fortunately, we were just finishing digging our holes in the sand along the river when the rounds started landing in and around us. My men immediately dove into their holes as the first of twenty-two 8-inch artillery rounds tore into my Company position. My "personal bodyguard" had dug part of the hole I was going to share with him, and I was digging in the hole when the barrage started. I had just told him a few minutes before to monitor my radios that were lying by our hole while I dug. After the first few rounds hit, I suddenly realized he was not in the hole with me. I quickly looked over the edge of our hole and saw him sitting on the ground by the radios. I yelled, "Jenkins, what are you doing?" He yelled back, "I'm on radio watch, Skipper!" I shouted for him to get in the hole with me. When he jumped in, his huge boots and large body frame landed on my back, and I thought I was going to be a casualty.

When the "incoming" finally stopped, our Arty Officer told me the problem he had while trying to get the bombardment on us to stop was that he was not sure what artillery unit was firing the rounds. Just as I had experienced a few months previously, he did not know who to call to stop the shelling of our position. The strategy of having the closest artillery unit that was available respond to a fire mission request often produced a faster response…but if something went wrong, as it did in this case, it became an unforgettable nightmare because we were not sure who was firing the rounds. Other than some "busted eardrums," we survived someone's mistake. I believe Divine Intervention played a role in our digging our holes in the soft sand because the rounds penetrated the surface before exploding, and it limited the injuries to my men.

In yet another bizarre turn of events, my Marine who had been sent to the Brig for hitting his squad leader several months earlier had just returned to my Company the day before this accidental shelling by friendly fire. He had an entirely different attitude and he appeared to be a "squared away" Marine. Unbelievably, he was one of the Marines whose eardrums were ruptured. He was medevacked and sent back to the world as a result of his injuries.

My personal bodyguard needs some explanation. A young Marine Private named Jenkins had recently been assigned to my Company. Shortly after first light, following his first night with us, a Corporal from one of the platoons literally dragged Jenkins over to my hole. Jenkins had a split lip, and it was obvious he had been punched in the face and eyes. His nose was bleeding, and he was sobbing. The Corporal said they had found him sleeping while he was supposed to be on guard. I told the Corporal to leave him with me.

After a minute or two, I asked Private Jenkins what had happened. He admitted that he had been asleep and went on to tell me the Corporal had threatened to kill him if he ever did that again. I tried to explain to him that he had jeopardized all our lives. I also told him that the Uniform Code of Military Justice, or UCMJ, as I understood it, possibly allowed for the death penalty for his offense in a war zone, in certain cases and under certain circumstances. I hoped that the explanation would impress upon him the seriousness of his "sleeping on post."

I visited with him for a while until he calmed down. To encourage him, I told him I would keep him with me for a few days, and I asked him if he would like to be my "bodyguard." He said he would. One of my troops was carrying both a M-16 and one of our Winchester 12-gauge shotguns. I had that Marine give the shotgun to Jenkins to carry. It was almost comical, other than some of the obvious inconvenience it caused me. Jenkins would follow me so closely that when I stopped for any reason, he would sometimes bump into the back of me. My Radio Operators continuously had to tell him to get out of the way so they could get a handset to me for communications, but he was making every effort to do exactly what he understood he was supposed to do. This

Marine stood probably 6'3" - 6'4" tall, and easily weighed 220 lbs. He had size 14 feet, and I later had considerable difficulty getting jungle boots that would fit him.

I tried to learn as much about Jenkins as I could so I could find him duties that would provide essential services for us, while at the same time allowing him to realize some success. I learned he had never seen his parents. He had been raised by an older sister, and it had taken him eight months to finish boot camp. That was an extraordinary amount of time for his initial training. He had an enormous interest in mechanics, and from that time forward, I gave him every opportunity to be involved in anything that was mechanical, even though it was very limited in a rifle company. He made every effort to do things right, and when I looked over the edge of the hole and saw him sitting by the radios during the 8-inch artillery barrage, I can assure you he had the shotgun right beside him.

Some of my proudest moments were watching Jenkins being transformed into a Marine who felt self-confident and one who believed he was part of the unit when he rejoined a platoon. He became a Marine that could be responsible and "hold up his end of the deal." On my last day as the Commanding Officer of Echo Company, I promoted him to Private First Class, or E-2. I have thought of him many times through the years and I have always wondered how he progressed through life, both during and after his service to our Country and the Marine Corps.

Overall, Operation Ballard Valley led by Col. Mueller was mostly uneventful from the perspective of initiating enemy contact, and we concluded it on May 3rd. The next day, my Company was chosen to be one of two companies to initiate what became Operation Allen Brook. The Operation began on May 4th, and a few days later on Mother's Day, 1968, we became engaged in a battle the Marines in my Company will never forget.

Operation Allen Brook

COL. MUELLER NOTIFIED ME that I would next be taking my Company on a search and destroy Operation on the Go Noi Island north and east of An Hoa, close to where we had just completed Operation Ballard Valley. He did not identify the Operation by any particular name. Again, it was the area west of Chu Lai the ROKs (Republic of Korea) had been assigned to defend a number of years before. As I understood the briefing, it was in the same area I mentioned when I described my amazing flight in the backseat of the F-4 Phantom jet. (See **A Dream Came True** chapter.)

The practical result of the anti-war political pressure in the ROKs' home country had ultimately allowed the area in South Viet Nam that they were responsible for defending to become an "R&R" area for the NVA and VC. The enemy forces were operating in that area with relative impunity and the "safe haven" aspect of the area had become much more obvious after the Tet Offensive. My Echo Company was to be one of two Companies to initiate the Operation, and military records indicate that the other Company was Golf Company, also from Second Battalion, 7th Marines.

In preparation for the Operation, my Company had been dropped in by helicopter north of the Liberty Bridge. It was predicted that the landing site was far enough from An Hoa to deny enemy intelligence sources any indication that we might be preparing for an Operation in the An Hoa area. As we moved overland south toward Liberty Bridge from the landing zone, we came to a small Marine fortified position and we stopped there for a while since we had some time to spare.

I introduced myself to the Company Commander. The Captain in command of the unit was Charles "Chuck" Robb, the relatively new son-in-law of President Lyndon Johnson. We visited for a couple of hours, and he aired his frustrations with his limited duty because of the enormous problems that would be generated if he were to be captured or killed. He appeared to be a focused Marine Officer, despite all of the distractions created by his familial connections. I later watched his political career develop in the United States Senate from afar, but I never had the opportunity to visit with him again. We left his unit's position later in the afternoon and moved to the river on the north side of Go Noi Island where I attended a briefing that night.

The briefing was conducted by an Officer I did not know. We were informed that Go Noi Island was possibly the base for the Viet Cong's Group 44 Headquarters, and military records now indicate that the enemy units in the area included the R-20 and V-25 Battalions, and the T-3 Sapper Battalion. In addition, these current records indicate that elements of the People's Army of Viet Nam, or PAVN, 2nd Division were also located on the Island. There seemed to be several "unknowns" in the briefing. Hindsight being 20/20, I cannot imagine why two rifle companies totaling approximately 350 Marines were being sent in to attack a stronghold of what could easily have been over 1,500 enemy forces. However, the Marine Corps "motto" then was, "Ours is not to wonder why, just to do or die."

In the briefing, launch time was set for 0200 hours, 4 May 1968. It became readily apparent that the supposed search and destroy mission was quickly becoming something much larger and significant. The order of operation was that we would be crossing the Thu Bon River on foot under cover of darkness. We had to ford the river because the Liberty Bridge had been destroyed by VC and NVA sapper units. The river was on the north side of the Island and flowed eastward toward the ocean a few miles away. Maps indicated that once we crossed the river, there would be a road on the right flank of our initial direction of movement to the south of the bridge. Briefing intelligence also indicated that beyond the right side of the road was a large rice paddy area approximately one-quarter to one-half mile across. According to the briefing, the paddy area was shaped like a horseshoe, with the open end of the "shoe" facing the road. There was a tree line along the edge of the paddies, and intelligence sources indicated that it was heavily fortified. The absolute orders were that, even though I could expect sniper fire from across the large rice paddy area west of the road, no matter what happened, my Company was not to cross the road unless ordered to do so.

The direction of the "search and destroy mission" was to parallel the road in a southerly direction and then to begin turning eastward with the edge of our left flank being the Thu Bon River. We were told that we could expect to find large amounts of supplies and armaments that would most likely be defended by enemy forces at all costs. Additional intelligence indicated that everyone in this area of operation was considered to be enemy combatants, regardless of who they were, and we were not to take any prisoners. In essence, that meant to kill everyone who was there.

I was to have five tanks to use as reinforcements. I posed a question at the end of the briefing and asked if we could have a chance to recon the area at first light and initiate the attack later in the morning after we had a chance to at least view the terrain. I thought it would be prudent military practice since the intelligence indicated that we could expect the extensive use of booby traps and land mines. I believed that it was in the best interests of my troops to do so, but my request was quickly and adamantly denied.

Historical documents indicate that the Third Battalion, Seventh Marines (3/7) had been in this general area the previous month on Operation Jasper Square. At that time, I thought that could have been when the abandoned Marine we found had been left behind. I discovered later that my assumption was incorrect.

Since the bridge over the Thu Bon River had been destroyed, ground transportation across the river could only be accomplished by the use of a pontoon bridge, if it were ever able to be constructed by U.S. forces. We launched at exactly 0200, and in the darkness, I located what I thought was the road we were to follow. We moved very slowly because of the possible mines and booby traps. We were on the east side of the road across from the large rice paddy area we were prohibited from entering.

As the darkness began to fade into dawn, we began to receive sniper fire from the tree line on the other side of the paddy area, but it was not accurate. It appeared to be the kind of harassment that the briefing had described. Instead of pursuing the sniper fire, I stuck with the Operation Plan and followed the Orders of Operation we had been given during the briefing. I began turning my Company slowly to the east away from the large rice paddy area. There was thick underbrush with some light enemy resistance. Our excellent tank crews in one of the tanks took out a machinegun placement with a 90 mm round. However, because the machinegun and its crew were so close to the tank, the delaying mechanism on the round that prevents it from exploding too near the tank caused the round to create a hole in the underbrush where the machinegun was located. Pieces of metal and enemy bodies flew through the air, and we heard the tank round explode a second or two later some distance away.

As we broke out of the brush into a small clear area, some women and children ran out of the brush on the other side of the clearing and stood in front of us. I quickly notified my troops not to shoot them. It appeared to me that it was a tactic by the NVA/VC to slow our movement, and it did. I immediately had to make a decision. As a Marine Officer, I had no doubt that I was to follow orders. However, in this particular case, my personal values instantly inundated my mental vision of what decision should be made, and for what reason.

With the women and children huddled in front of us, I called back to the Command Center and told them that we had encountered apparent non-combatants, and I asked them what they wanted me to do with them. At that point, killing everyone in the area of the Operation was not an option as far as I was concerned. After a short time, I received a message that we were to take these non-combatants with us. The only thing we could do was to place them on the tanks. It was very dangerous because the turrets could not be moved or operated without injuring or killing the people sitting on the tanks. In approximately an hour, we had 40-50 women and children riding on the tanks. The steel plates covering the engines in the rear of the tanks were becoming very hot from the intense jungle heat and large diesel engines beneath them. The children were not wearing any clothing, and they began to scream and cry because their bodies were being burned by the hot metal.

If experience had taught me anything, it was that the NVA and VC would attack those tanks killing all of the women and children, if it provided them the opportunity to destroy the tanks. I notified our Command that the situation was critical, and the transportation of the non-combatants on the tanks was jeopardizing my mission. After a few moments, I received a message saying to release the women and children, and they disappeared quickly into the underbrush.

Historical military documents of the 3rd Battalion, 26th Marines now indicate that again, my reliance on my personal values instilled in me by the teachings of my youth and reinforced during my college education at John Brown University had an enormous humanitarian effect. Those records indicate my decision not to kill innocent non-combatants led to approximately 225 civilians being evacuated from Go Noi Island on the first morning of the Operation, even though everyone was supposed to have been considered enemy combatants and killed. I have thanked God many times for having given me the courage to make the correct decision that day.

We could tell that we were being "stalked" by enemy forces, but resistance was light. As we moved eastward, the river we had crossed became our left flank. I tried to ensure that our formation was well defended as we moved along the river, which is somewhat difficult to do when you are trying to conduct an Offensive Operation. However, it became very clear that we were being "tolerated" by the NVA and VC, and that they would attack us when they believed they had the best advantage.

On the second night, I believed the position where we dug in gave us a good defensive position. Keeping the size of the holes as narrow as possible always decreased the possibility of being hit by incoming mortar or rocket fire. We made it a practice to allow people who were in individual holes to lay on the ground in a fetal position over their hole when it was their turn to sleep, so that if anything happened, all they had to do was straighten their legs and they would fall into their holes.

The NVA and VC were amazingly accurate with the 60 mm mortar. They could quickly set the tube on the ground and simply hold it at the proper angle to obtain the trajectory of the round needed to hit where they wanted it to. On that night, we suddenly heard the "thunking" sound of mortar rounds being dropped into the tubes. From the sound, we could tell they were close to us, and it was obvious that the rounds were most likely being fired toward our position. In seconds, we could hear the mortar rounds coming through the air, and just as several of us shouted, "Incoming!", the first of about ten rounds of mortar fire struck our position. All of them landed within an area about 50 feet square, which was where my Radio Operators were located near Sid and me. Amazingly, no one was killed, but three of my Radio Operators were wounded. Sid and I were lucky and just got our "bells rung." The result of the mortar attack was that we needed a night emergency medevac, and we had to get three troops, without any Radio Operator experience, to become reliable Radio Operators in a matter of a few hours.

As I have said, Radio Operators are a special breed. One of the replacements was a "gung-ho" Marine named "Gabe" Quinones. He was fearless, as good Radio Operators had to be, and he immediately became my Company Radio Operator. Cathi trained him just before being rotated back to the world, so it was a logical step for him. Cathi had recommended him to me without reservation. The fate of the entire Company often depended on the Radio Operator's courage, self-sacrifice, and ability to keep their senses when chaos was erupting all around them. They had to decide when I needed to answer a call and prioritize various calls. I could always ask Gabe for his opinion without him thinking I was indecisive. On Mother's Day, 12 May 1968, his courage and ability to function without fear negatively affecting his sound judgement was absolutely essential to the survival of our entire Company.

When we left our defensive positions, we always tried to fill in our old holes. The holes we would leave behind were prime targets for booby traps because, all too often, troops would go back into the old holes because it did not take so much digging. We constantly reminded our Marines not to do that. In the early evening a couple of days later, a very thick fog moved onto the coast from the South China Sea. We could not see our hand in front of our face. One of my men accidently stepped into a partially filled hole that was booby-trapped. It severed both of his legs near the knees, but because of the relatively "clean" cut of the bones and tissue, our Corpsmen were able to dramatically slow the loss of blood with tourniquets. I immediately had my air Radio Operators call for a medevac, but we were told the weather conditions prevented us from being able to have my wounded Marine lifted out. As we often did, I asked the Radio Operator to "dial up" on the aircraft frequency and to begin randomly calling for help. After a short time, he received an answer and he handed me the handset.

I was talking to an Army Warrant Officer who had just lifted off in DaNang, and I explained our problem to him. He instantly said he would try to make the medevac and he asked me where we were located. I told him my general coordinates, but with the inaccuracy of the maps and the fog, we both decided we would have to use sound to make some of the determinations. It was not too long until we could begin to hear a chopper. I talked with him until it sounded like he was above us, and then he hovered and began to slowly descend. Soon we could begin to feel the "prop wash," and a few short minutes later we could almost touch the chopper before we could see it, and then all we could see was the dim red lights on the instrument panel of the aircraft. As you can imagine, he was not using any other lights out there in the "bush." We loaded the wounded Marine onto the chopper, and he quickly vanished into the intense darkness.

The Army had a Warrant Officer Helicopter Pilot program in which 19-year-old Army soldiers could become pilots, and this was one of them. The heroics of these young men were amazing to me, especially because I had become a licensed private pilot while I was stationed in Virginia before going to South Viet Nam. I understood, to a limited extent, the danger involved in what this young pilot was doing. He risked his life for fellow Americans he had never heard of before. As I thought about how thankful we were for him that night, I was reminded about the old saying that, "There are old pilots, and there are bold pilots, but there aren't many old, bold pilots." I hope that young pilot was the exception to that "old general rule."

Parts of the Marine Corps tactics were that units like mine would be used to bait the enemy forces to expose themselves by attacking one of our units with a superior force. If everything worked as planned, Marine forces would then react with superior fire power and reinforcements to assist the unit being attacked after the enemy had tactically exposed themselves. That was part of the tactics for this "search and destroy mission," which the 3/26 records indicate was initially dubbed the "No Name Operation." The size of the enemy resistance that we could sense around us was alarming.

Historical documents also indicate that Golf Company 2/7 was withdrawn from the Operation on 7 May for unknown reasons, and it was replaced by Alpha Company of the First Battalion, Seventh Marines (1/7), but I was unaware of the changes at the time. I did not have any contact with the other Companies operating on the Island during what became Operation Allen Brook. The only tactical communication I had was with the Central Command of the Operation.

As we continued eastward along the river on 9 May, we came to the defunct railroad track that ran from Hanoi to Saigon. It was a single track and it did not appear to have been used for several years. Intelligence indicated that there was not any enemy activity east of the track. However, the track bed was elevated so high and the sides of the bed were so steep that it was impossible for tanks to cross it.

To get to the other side of the track, the tanks we had with us would have to go south about one-quarter of a mile along the track bed to get under the track where there was a railroad bridge over a stream. As we approached the tracks, we passed through an abandoned village that left an open paddy area between us and the railroad track.

The tanks had just started to move south toward the railroad bridge. To secure our advance and to be able to provide some protection for the tanks, I moved my Company past the edge of the village. We began to move across the paddies toward the track bed in a sweep formation that extended from the village to the railroad bridge. When we were about halfway across the paddy area, the gooks attacked us from the elevated railroad track bed. They had us pinned down with no protection except the paddy dikes.

Because of our position in the paddies and the enemy positions on the opposite side of the tracks, the tanks were not able to assist us. They were caught in a very difficult defensive position also and were dangerously exposed to RPG and B-40 rocket fire, especially since they could not go up and over the elevated tracks. We were able to prevent them from launching a frontal attack on us and destroying the tanks by keeping concentrated fire on the track area while I called for air support. An airborne Forward Air Controller, or FAC, controlled the close-air support for me, and he had the first sortie of fixed-wing aircraft there within about 15 minutes.

A FAC was one of those brave individuals who flew his own very small aircraft loaded with several radios. He communicated with ground units like mine while directing fighter/bombers and helicopters in support of ground units. Most of the aircraft the FACs flew were not much larger than a Piper Cub. Nearly all the ones I saw were single engine, and they always flew low and slow. They established the locations of the ground units and communicated directions to the fixed-wing aircraft. Some had the ability to fire small rockets that could mark targets with smoke. That was of great assistance to us, because on the ground it was sometimes impossible to mark targets in a way that could be seen from aircraft traveling at 300 to 500 miles per hour. I was always amazed that a small plane of that size and speed could avoid being shot down, but the FACs were always very courageous and resilient.

The enemy forces that day made a huge tactical error by using the railroad track as a position because the aircraft could easily see the tracks from the air. In a series of air support attacks that lasted over two hours, Marine Corps, Navy, and Air Force aircraft dropped bombs and napalm so close to us that some small pieces of napalm canisters landed inside our position. It was an unbelievable display of fire power that was so accurate that even a difference of 50 feet could have caused numerous casualties in my Company. For that outstanding expertise, I will be eternally grateful. The GPS and "Smart Bombs" were not in use, or maybe even heard of, in the Viet Nam Theatre. However, aircraft traveling at 300-500 mph dropped ordinance with deadly accuracy numerous times that day. Enough cannot be said about the professional skills of American pilots in all branches of the military, and their outstanding abilities are yet another reason I am able to write this book.

I was never sure why the Commanders of the Operation wanted me to move my Company to the east side of the railroad, but the next morning, 10 May, we moved south to the railroad bridge location and moved the tanks under the tracks to the east side. After digging in around a hamlet and spending the night at that location, we conducted offensive search and destroy operations that resulted in limited contact on 11 May. Even though we were suffering some casualties, both wounded and/or killed every day, our contact on that particular day had been dictated by the probing and "hit and run" tactics of the superior enemy forces.

"Mother's Day Massacre"

THE NEXT MORNING, I was told to move my Company back to the west side of the tracks where we had been just two days before. It was Mother's Day, 12 May 1968. I had an ominous feeling that what we had just been ordered to do was not good operationally or tactically. The previous attack on my Company on 9 May from the railroad track location was an indication of what the enemy forces were willing to sacrifice to defend this area. The losses they sustained from our air strikes that particular day had to have been staggering. However, they seemed willing to "pay any price" to defend this area that was apparently of such importance to them. For whatever reason, the "No Name Operation," started by two rifle companies, had become Operation Allen Brook, and its early records were to include the tragic events of Mother's Day, 1968.

We approached the bridge to get the tanks back under the railroad track to the area where we had been just two days before. I told the Tank Commander in charge of the tanks that I was worried that the area could be mined. The gooks knew that we were forced to go under the bridge to get the tanks to the east side of the tracks...and they obviously knew that it was the only way for us to get them back to the west side of the tracks. When he questioned me about the wisdom of the new order, I explained to him that our Commanders told us to return to the west side of the tracks to operate for the next few days.

I called the Operation Command and requested that a mine detector be sent out so we could check the area under the bridge before the tanks moved back through to the west side of the tracks. I was told none were available, so our only option was the rudimentary process we had been taught in Basic School which involved using our K-bar, or large knife, we carried to very gently probe the ground to see if we could possibly detect any sort of metal, which could be a land mine. We tried that but no one found any metal close to the surface of the ground.

The first tank went under the bridge without incident. The second tank struck a mine, and it blew the track apart and damaged some of the track idler wheels. Fortunately, it did not penetrate the fuel cells in the bottom of the tank or seriously injure any of the tank crew. Unfortunately, the space under the bridge needed to move the tanks was now blocked. The first tank that made it through safely backed up to the disabled tank and the crews attached a tow cable to the "dead" or "blown" tank. They were able to pull the blown tank on through to the west side of the tracks and out of the way.

The Tank Commander assumed that with that mine detonated, that was probably all the mines. When the third tank went under the bridge, it also hit a mine resulting in similar damage to the tank and crew as had been suffered by the tank in front of it. While they were pulling the second damaged tank through the opening, I was able to reach Operational Command again to find out what they wanted us to do with the blown tanks. I was told that it was necessary for us to tow the tanks with us as they could not be abandoned in the field, and a tank retriever could not reach the damaged tanks in that location. I reminded them that if either of the two remaining tanks hit a mine, we would not be able to tow three dead or blown tanks. It appeared that there was not anything else we could do, so we had to take our chances. The last two tanks moved under the bridge successfully without any further damage occurring.

It was approximately 1000 hours. I was ordered to move my Company west, paralleling the river, over a large paddy area approximately a mile long and one-quarter to one-half of a mile wide. The river was now to be our right flank. Two of our tanks were towing the two disabled tanks which rendered them essentially useless except for their turret. Without their maneuverability, they were "sitting ducks" for VC and NVA RPGs and B-40 Rockets. In addition, they could only offer very limited barrier protection for ground troops. Along the south side of the paddies was a tree line and hedgerow that curved toward the river on the west end of the paddy area, but the tree line did not go completely to the river. Nothing about the situation felt right to me or anyone else in the Company who had command responsibility.

As we began to move westward away from the railroad track, I had troops from the platoon led by 2nd Lt. Jackson on the left flank of our formation moving through the tree line on the south edge of the paddies. The rest of my Company was spread across the paddies to the river, which was our right flank. We were moving in an on-line sweep going west, and we had moved at least three-quarters of the distance of the length of the paddy area. The pace of our sweep movement was determined by the progress of the platoon in the trees on our left flank.

There was a FAC in the area, as one had been the previous days. He called me and asked me if I had troops in the tree line on my left, or south, flank. I replied that I did. He said that I should be prepared because from his vantage point, he was sure the enemy forces were going to "hit" us from the south on our left flank. He went on to say that he had "never seen so many gooks in the open" since he had been in-country. Just as he finished that statement, the NVA and VC attacked our left flank with a staggering volume of small arms fire, RPGs, and B-40 rockets. Immediately, all my troops in the paddies were pinned down behind the paddy dikes.

I was in the middle of the formation in the paddies with the two tanks being towed behind me and the single "maneuverable" tank in front of me. The sudden small arms fire was so intense, it was difficult to roll over so I could get a better view of the left flank. My Company Radio Operator Gabe handed me the handset and he said it was Lt. Jackson, the PC of the platoon on the left flank. Lt. Jackson told me I needed to get there as quickly as I could because he had been wounded and the gooks were coming out of the trees after them. I quickly called the Tank Commander, or TC, and told him that we needed help on the left flank. I told him I was behind him and to his left, and that when I got to the tree line, I would let him know where all my troops were. The tank 90 mm weapons could cause casualties among my Company when they fired, especially if the TC did not know exactly where we were. I called the FAC back and told him to use his discretion to try to keep the gooks off us as much as possible.

I told Gabe to follow me, and we started across the paddies to the left flank. Gabe never hesitated, and he knew we were facing overwhelming odds. No matter where I went, he was always there, and I know he answered calls and made decisions that I could not because of the chaotic conditions. Even from the distance I was from the tree line on the left flank, I could readily see Lt. Jackson's platoon was in extreme peril as the NVA and VC forces were trying to overrun them.

Somehow, Gabe and I got to the tree line where the Jackson Platoon was trying to hold off the enemy assault. I saw Jack, who was Acting Company Gunnery Sergeant that day, maneuvering toward us to help Lt. Jackson's platoon, but he was still several paddy dikes away. Somehow, Gabe and I had covered the 150-200 yards quickly through the withering small arms fire and rocket explosions.

We found Lt. Jackson with a leg wound and several other wounded Marines. I radioed the TC, who had turned and was following me toward the beleaguered platoon on the left flank, and I asked him if he could see me. He said he could, and I told him one of the gook machinegun placements was just to the right of where he could see me. That gook machinegun position was pouring fire into the center of my Company and had them pinned down in the middle of the rice paddies in the area where Jack was. I told the Tank Commander I would try to let him know where all my men were as quickly as possible, but that he had to take the machinegun out regardless of where we might be.

I was aware of the danger of firing the 90 mm too close to us, but it was a matter of where we were going to possibly lose the largest number of troops. I had seen immense damage to vehicles parked beside and behind artillery pieces when they were test-fired. This tank 90 mm was much closer than that to us, and I was the closest one to the tank. We were literally being chopped up by the enemy machinegun that was closest to me, so I told him to fire. He placed a 90 mm round dead center on the enemy position and the instant effect was devastating to the attackers as it obliterated the weapon and the enemy gunners there.

The tank round blast was so close to us that it caused tremendous shock, concussion, and pain to those of us in front of the tank. My vision was lost for a few seconds as my sight became horizontal lines, like that of the first black and white television sets when they would lose focus. The pain in my legs and arms was tortuous, and when I tried to get up to attack into the tree line with my troops, my equilibrium was seriously affected. I had trouble keeping my feet under me. The force of the tank blast was so horrific that it seemed to shock the enemy attackers, causing them to retreat back into the trees for a few minutes.

This gave us time to go into the trees after them to try to provide time and cover for Jack and others to begin dragging and carrying the wounded away from the left flank and north toward the river. I believed the only chance my Company had to survive was to consolidate into a defensive position with our backs to the river. Unbeknownst to me, Sid had already started organizing the defensive position.

There is not any question in my mind that the suppressing fire from the tanks saved our lives because of the overwhelming number of enemy attackers who were trying to charge out of the trees. The effect of the tank blast on us physically was somewhat similar to the effect of shock waves in the ground from bombs that were dropped close to us when we had to use close-air support. The effect on the enemy attackers was much more devastating. As we went into the trees after the initial attackers, another machinegun opened up just to the right of my face and some of my hearing seemed to instantly fade away. We were able to knock out that machinegun and force some of the closest gooks farther back into the trees. The shock from the initial 90 mm round seemed to have created confusion in the gooks, possibly because they did not think we would use a tank round that close to our own troops. It could also have been because these particular enemy troops may have never had a tank used against them, and they may not have been aware of the firepower the tanks had. Whatever the reason, it gave us the advantage for a few minutes.

Jack began carrying wounded Marines to a partially protected area where the tree line curved toward the river. After the tank had taken out the machinegun position to my right, my Marines in the middle platoon of our Company formation who had not been wounded were trying to move toward the left flank where I was attempting to help reinforce Lt. Jackson's platoon. The Tank Commander had ordered his other two operational tank crews to unhook from the tanks they were towing, and they began moving toward our left flank. After the shock of the initial 90 mm round wore off, the gooks quickly counter-attacked through the trees, and they moved up more anti-tank weapons and reinforcements. They launched multiple B-40 rockets and RPGs at every conceivable location on the tanks trying to disable them.

The middle platoon's reaction, combined with the 90 mm rounds and .30 caliber machinegun fire from the tanks, kept the enemy forces pinned down and inflicted heavy losses on the attackers. We were able to begin withdrawing toward the defensive position so the river would be to our backs, dragging and carrying our wounded with us. While we were trying to consolidate my Company by pulling everyone back to the defensive perimeter bordering the river, the three tanks were leveling devastating .30 caliber fire into the tree line. Once they could see we had all our people pulled back far enough so that we would not be injured by the tank 90 mm rounds and machinegun fire, they literally "put up a wall of steel" and explosive rounds to cover us.

I glanced over to where Jack had carried the wounded, and it seemed he had a lot of men there. There was an opening in the tree line, and it appeared to me that the gooks were trying to move the attack to our right flank along the river. If they succeeded, it would have allowed them to reach the position where Jack had carried the wounded Marines he had rescued. I grabbed a Marine that was involved in trying to get everyone pulled back to the river and told him to go to where Jack was, and tell him that the gooks were trying to "flank" us on the right and they were trying to get to the river. I also told him to tell Jack that it looked like the gooks were trying get to where he had the wounded men. As we kept withdrawing and consolidating toward the river, I saw Jack beginning to carry the wounded from the first place he had taken them to the medevac position Sid was establishing.

Sid was on our right flank when the attack began. Without any communication with me, he instantly evaluated the situation, and in an outstanding display of excellent Marine Corps tactical leadership, he quickly began to organize a half-moon defensive position with our backs to the river. If he had not done so, we would not have been able to have an organized defensive position to withdraw to. The casualties were overwhelming, and it was only because of his excellent leadership that we were able to begin the evacuation of troops so quickly. When I reached the defensive perimeter, I told him to handle the evacuation of the medevacs as his part of the heroic efforts of all my men to survive the withering attacks.

By this time, the heat was overwhelming, and I had difficulty keeping my vision clear, as well as feeling extremely nauseous. Jack was bringing what turned out to be nine wounded Marines to our new defensive position. They were the same Marines he had previously carried back from the left flank earlier. I could see that he was near exhaustion. However, no one else could help him because of the enormous casualties we were suffering. When he brought the last of the nine wounded Marines to our position along the river, he collapsed on the ground and seemed to be having a convulsion. I thought for a second that he had been wounded, but then realized he was probably suffering from heat exhaustion. I had two Marines grab him and drag him into the river where they submerged him, except for his head, and continuously poured water over his face.

Withdrawing our wounded to our defensive position allowed the medevac choppers to be able to finally get to us. From the surface of the water in the river to the top of the bank was about 10-15 feet. I asked the pilots if they could fly up the river from the east staying below the banks as much as possible and then come up over the bank at our position with the ammo and water. I told them we would load dead and wounded in the choppers before they dove back over the edge of the bank for the return back downriver toward the ocean. They said they would try...and as always, they "got 'r done." It was so crowded along the river for a while that one chopper landed on a Marine's leg causing a compound fracture. We threw him on the chopper, and he was flying away from where he was injured before the extent of the pain actually hit his leg. Each time the choppers came up over the bank and were on the ground, it was only for a minute or two at the most. Every second counted because enemy fire from all kinds of armaments was shot at the choppers when they rose up over the riverbank after they had "skimmed in" just above the water to get to our position.

The three movable tanks had gradually pulled back offering protection for us as we moved the casualties back and provided protection for the choppers when they came up over the bank. At one point, one of the tanks backed into a bomb crater and was stuck. The bottom of the tank where the fuel cells were located was exposed to the enemy position, so the Tank Commander moved his tank in front of the one that was stuck to protect it and to try to pull it out of the bomb crater. He crawled out of the turret, removed the tow cable from his tank which normally takes two men to handle, hooked it onto the tank that was stuck, had his tank pull the other tank out of the bomb crater, unhooked the tow cable, threw it back on his tank, and crawled back into his own tank. The Tank Commander did all this amidst enormous small arms fire and repeated attempts by the enemy to hit both tanks with B-40 rockets and RPGs.

We were not aware that a hospital ship happened to be just offshore from where we were, and many of our casualties were aboard that floating miracle in just a matter of minutes. It will never be known how many lives that "coincidence" may have saved. I checked on Jack and I could not believe he was already walking around trying to help. At about the same time, I had a call on the radio that one of the tanks had stopped moving. I asked if the tank had been hit. They said they did not know, but they thought they heard a voice transmission coming from it.

After what I was dealing with concerning the burial issue, there was no question about what should be done. I told Jack that we thought we had heard a radio transmission from the tank sitting in the middle of the paddies, and that we needed to try to get to the tank. He jumped out in front of our perimeter and yelled for others to help and give him covering fire. He ran to the tank, crawled up on top of the turret and opened the hatch. He said later the tank crew had made the ultimate sacrifice, and just as he started down off the rear of the tank, a mortar round hit the top of the turret. The blast blew him off the tank and knocked him unconscious. The gooks were firing a large number of mortar rounds, and while they did not damage the tanks, when they hit on top of the tanks, they caused explosions in the air that could kill and wound Marines around the tanks. When Jack came to, other Marines were on their way to help him back to our perimeter, but he refused their assistance and made it back on his own.

The entire battle lasted about four hours. During that time my Company sustained 52 casualties, and the tanks attached to us had four killed in a tank that was destroyed by a B-40 rocket. For nearly 40 years I believed that the majority or all those casualties in my Company had died from their injuries. It was not until I went to an Echo Company 2/7 Reunion that I learned that only two actually died that day. The sacrifices of the tank crews, the bravery of the our Corpsmen who rendered outstanding initial medical care to my Troops with complete disregard for their personal safety, the courage of chopper pilots and flight crews, the heroic efforts of Sgt. Jack Johann, and the availability of the professional hospital ship staff so quickly proved to be the turning points in the lives of at least 50 of my men. In addition to that, every member of my Company displayed extraordinary courage and bravery throughout the battle. The heroic efforts of all my men risking their lives for me and each other upheld the highest traditions of the United States Marine Corps.

I nominated the Tank Commander to receive the Navy Cross for his heroic efforts that day, but I did not know his name. I have never seen any evidence of who it may have been and whether he received it. The reason I do not know his name is because I would write a very brief message of recommendation on a small piece of paper and send it back to 2/7 Headquarters. That would usually be the last I would know anything about that particular recommendation. Without the tanks and their respective crews on Mother's Day, 1968, our Echo Company casualties could have easily defied all estimates. The wall of steel the tanks poured into the gook positions allowed us to consolidate our defensive position along the river and very possibly saved the lives of our entire Company. Military records indicate that enemy losses were 80 KIA, and the number of enemy WIA is unknown. However, I do not know how the number of enemy KIA was determined.

I also nominated Jack to receive the Navy Cross. I believe Jack was solely responsible for saving the lives of nine Marines that day. His bravery, tenacity, and leadership exceeded the highest standards of conduct of the United States Marine Corps and the United States Navy. Sadly...unbelievably...Jack's award was downgraded to that of a Silver Star by someone who was not even there. It was an arbitrary decision for which no one had to give a reason. I had been in the Marine Corps for such a short period of time, I did not know "what kind of games had to be played" or "certain words to use" when nominating someone for an award. After years of reading about military awards, I realize now I should have nominated him to receive the Congressional Medal of Honor.

Sid was also amazing during that battle. He and I thought so much alike about so many issues that I trusted him to do what had to be done, and I believe he always knew I had his back. He was "everywhere" on Mother's Day, 1968, but I seldom caught sight of him in the midst of the horrible chaos. Without him there, only God knows if we would have prevailed, or how many more mothers would have received notification that their son gave his life for our ungrateful Country on the day for which she was to be honored. Sid knew what had to be done and reacted accordingly. We coordinated almost psychologically that day because we both knew what was needed, and he did not have to be directed by me. Sid is the epitome of a great Marine leader, and even a better person.

When a Company loses 52 members of its fighting force in one day, it becomes too small to carry out the duties of a tactically sound Marine Rifle Company. Echo Company had experienced casualties during the entire Operation prior to that point, and every person in the Company Office Staff located at the 2/7 Headquarters who could possibly come to the field had already done so. There was no one else left who could be brought to our location to help us carry out our mission. I had my Corpsmen evaluate all our "walking wounded," and I had to keep all my men in the field that I possibly could. In the highest traditions of the Marine Corps, not one of the wounded Marines objected.

However, the result was that some of my men did not receive their Purple Heart Awards because of the administrative difficulties associated with the timing of their wounds, and the tactical situation we were in. Lance Corporal Martin Bellman was one of my Marines who did not receive his award. After becoming aware of the error, three of my outstanding Echo Company Marines, Sergeant Kell P. DeVoll, Sergeant Joseph W. Freda, and Sergeant John F. Blaize, joined me in an attempt to correct this very unfortunate situation in 2018. Lance Corporal Bellman was promoted to Corporal before he left the Marine Corps. Almost 51 years after the "Mother's Day Massacre" on 12 May 1968, we received notice that USMC Corporal Martin Bellman would receive his Purple Heart Award.

My Company was relieved three days later at An Hoa on 15 May by another Company who waded across the river as we had. I briefed the new Company Commander as we stood along the south bank of the Thu Bon River. In absolute terms, I specifically warned him not to cross the road just south of An Hoa and not to go into the rice paddy area on the west side of that road. I told him he may receive sniper fire from there, but to ignore it. I found out later that before the day when I briefed the new Company Commander was over, his Company had gone down that road and they had received sniper fire from the trees to the west. Ignoring my warnings, he led his Company into the forbidden paddy area in pursuit of the snipers.

The source also said that within two hours, every Officer and Noncommissioned Officer in that Company was either killed or wounded, and that the devastation was so severe that not one round was being fired by the entire Company when forces arrived to reinforce them. My source also said he has confirmed that the entire Allen Brook Operation was suspended for a period of time after that. Military records indicate that what had begun as a search and destroy mission by two Marine Rifle Companies, one of which was mine, eventually became Operation Allen Brook. Parts of six Marine Battalions, which included as many as 24 Companies, were committed to the entire Operation.

Operation Allen Brook lasted until 24 August 1968 according to some historical documents. Records quoted in the documents indicate that 172 Marines were killed, and 1,124 Marines were wounded during Operation Allen Brook, while enemy losses were 917 killed and 11 were captured. The number of enemy wounded is unknown. It has occurred to me in later years that the "miracle" of the hospital ship being offshore so close to the Go Noi Island on Mother's Day, 1968 may not have been a miracle at all. It may have been there because Marine Commanders at higher levels had intelligence information that the size of the forces we faced was enormous, and they anticipated heavy Marine Corps losses. Whatever the explanation, I thank God every day for that floating medical facility and what it meant for the families and loved ones of at least 50 of my men on that fateful day.

R&R

BECAUSE OF CASUALTIES AND ILLNESSES, a few of my Marines who were able to stay healthy and uninjured were able to take two or three Rest and Relaxation, or "R&R," trips. Of course, they would have had to have the money for the trips, but it was possible. Unfortunately, several who I approved to take R&R did not make it past the departure facility where they would stay overnight before departing. The Red Cross organization provided young ladies to befriend those young men at the R&R Centers, and my Marines told me that there were several ways to spend your money there quickly. Some said they made the mistake of doing so. There were several R&R destinations, such as Tokyo, Japan; Bangkok, Thailand; Melbourne, Australia; and Hawaii. It was an excellent policy for our armed forces personnel in South Viet Nam. I took mine in the spring of 1968 to meet my wife in Hawaii. I had been notified that I was being sent to Camp Smith, Hawaii after completing my tour in VN. We met there so she could spend some time looking for a job teaching school during the 1968-1969 school year.

The United States military did everything they could to assist the personnel on R&R, but problems did occur. In our case, our planes landed at about the same time, so all the plans we had made to meet each other did not work. In anticipation of such problems, the military had note boards located at the airport and all the major hotels. If you were looking for someone, you left a note with contact information for the person you were meeting and you both had to keep checking in all the buildings until you finally found each other. In our case, it took about four hours. The three days flew past, and it was hard to believe we both had to return to our separate lives.

While I was there, I went to Tripler Army Hospital to see if any of my Marines who were wounded, or seriously ill, were there. I found the Marine who had picked up the booby-trapped hand grenade and threw it at the last second. (See **Lives Changed Forever** chapter.) He appeared to be in good health, but he was "asleep" when I stepped into the room where he was being cared for. The medical personnel told me that everything indicated a very small sliver of metal had entered his brain through his eyelid, rendering him unconscious. They reiterated what Sid had seen the day it happened and said there were no physical marks of any injury. While I was there, he stretched and yawned as someone would if they were awakening. The medical personnel said he did that regularly, but the problem was he consistently did not indicate any signs of actual consciousness. I asked them to notify me of any progress in his condition in the future, but I never did receive any notification. My prayer was and is that he regained consciousness in the future and was able to carry on a normal, productive life.

I also found the four Marines who survived the defensive fire by the tanks attached to my Company when the listening post was in the wrong position during Operation Ballard Valley. (See **Operation Ballard Valley** chapter.) The sight of them lying there with wounds that were still gaping holes of raw flesh was nearly unbearable. All but one of them were on respirators struggling to breath and speak, but they seemed to appreciate me spending time with them. The visit to Tripler reinforced the horrors of my Viet Nam experience. The rejection of those of us in the United States military by many Americans was obvious in Hawaii, and it further skewed my sense of personal value and reality.

On the return trip to South Viet Nam, I was seated next to a Marine F-4 pilot and, as usual for me, our conversation went to flying. I described my "hop" over North Viet Nam and the elation I felt after pursuing one of my life-long dreams. I asked about his experiences in the skies over VN, and for some reason I felt compelled to ask him if he had received any Distinguished Flying Crosses, or "DFCs". He replied that he had been nominated for one while flying cover for a ground unit who was retrieving the remains of a Marine southwest of DaNang.

I suddenly felt as if I had been hit by an enormous electrical shock. My mind exploded with a "flashback" of me covering the young Marine with dirt in that shallow grave in the bomb crater. I had a tremendous impulse to get away from the conversation and off the plane, which obviously was not an option at 35,000 feet above the Pacific Ocean. I sat quietly for a short period of time until I became aware that the pilot was asking me if "something was wrong." With my pulse racing, I asked him if he had been told to "bomb the bomb crater." Stunned, he asked me how I knew that. I replied, "Because you missed." As we stared at each other in disbelief, the horror of the burial engulfed me. When I finished telling him what I knew about what had happened, he sat there in "stone-cold" silence. After what seemed like an eternity, we began conversing again, but the subject of the retrieval of the Marine was never mentioned again.

While the R&R trips were amazing, a glitch supposedly occurred with one R&R flight during that timeframe. Transcontinental flights in those days used land and water-based navigation stations for transoceanic flights. It was reported that one R&R flight had somehow missed Hawaii due to navigation errors, and by the time the plane found Hawaii, it only had 20 minutes of fuel remaining. True or not, travel was much different in those days than it is now with the use of satellite telemetry and Global Positioning Systems, or "GPS". Our return flight was around, over, and through a typhoon. When we looked out of the windows as we approached DaNang, we could see rice paddies an estimated 500 feet below. Collectively, all passengers were hoping and praying that even the NVA and VC were trying to stay out of the torrential rain, rather than shooting at low-flying passenger planes.

Special Landing Force

WHEN I REPORTED BACK to 2/7 after my R&R, I was relieved of my command of Echo Company, Second Battalion, 7th Marines on 23 May 1968, a Company Commander tenure of 7 ½ months. Col. Mueller reassigned me to be the S-4 of 2/7 for a little more than two months until my rotation back to the world. Shortly after my new assignment, we were notified that 2/7 was going to become the Special Landing Force, or "SLF." The purpose of the SLF was to have a Battalion-sized force at sea that could be deployed from ships in a variety of areas rapidly, which was the more traditional mission of the Marine Corps. A "land-based Marine Corps" was essentially what our USMC role in South Viet Nam had become. To my knowledge, it was the first time in Marine Corps history that Marines had been deployed in that manner to that extent. In addition, no one anticipated that this land-based deployment was going to last for approximately 10 years.

In order to prepare for the SLF Deployment, it was necessary for 2/7 to load aboard a small task force of Navy vessels for transport to the Subic Bay Naval Base in the Philippines. The mission was to make repairs, retool, replace some weaponry, train troops on both land and aboard ship, and to bring troop strength to the tactical levels as prescribed by Marine Corps Regulations. As the new S-4, I shared some of the responsibilities of coordinating our loading of personnel and equipment with the appropriate commanding officers in the Navy fleet.

We were at sea for approximately three days before we reached Subic Bay. The ship I was on was a World War II vintage aircraft carrier that had been converted to be a Landing Platform Helicopter, or LPH. I could not believe what it was like to have meals served on dishes, hamburgers for snacks, and movies. During some moments of relaxation aboard ship, I wondered if I was in the wrong branch of the Navy...but then I would always remember why I was a Marine.

There was huge anticipation for Liberty Call in a small town called Olongapo that was just outside the Subic Bay Naval Base. A footbridge gave access to the small town, and the bridge spanned a stream of raw sewage containing untreated human and industrial waste as it entered the Subic Bay Harbor. Philippinoes swam in the sewage whenever people walked across the bridge. They had cone-shaped containers in their hands, and they would ask us to throw coins to them. When a coin was thrown, there was a mad scramble to catch the coins before they landed in the incomprehensibly polluted water. At the end of the bridge was the epitome of racial prejudice that I had never seen before...a sign that said "Whites," with an arrow pointing one direction, and "Blacks," with an arrow pointing the other direction. I could not believe my Company had just spent months risking our lives for each other...living and dying side by side...and now we were being told by a foreign government that we could not go on Liberty Call together. To believe that the United States Government was not aware of this injustice seemed ludicrous to me. Apparently, those Americans in positions of responsibility chose not to do anything about the blatant racial prejudice that existed there.

Olongapo was nothing more than bars where prostitution was legal, and alcohol and drugs were cheap. Liberty Call presented serious challenges for some of our Marines. The natives knew that if they could marry a member of the U.S. military, they would be eligible to receive up to one-half of that individual's pay, depending upon their rank. In our briefings to Officers prior to our troops being released on Liberty Call, cases were cited where some foreign women were discovered to have married as many as six or seven military personnel. As a result, the massive military bureaucracy sometimes mistakenly allowed those women to collect one-half of each of their "spouses" pay checks from the United States Government. In real money, a Marine Private was paid $90 per month in those years. If they were or became married, he would have one-half of his pay, which was $45, automatically deducted and it would be sent to his "wife."

The only remedy we had available was to sternly brief our troops that if they took the vows of marriage while on Liberty Call, the marriage would not be recognized by the United States Government. We also warned them that if they became a new "groom" on Liberty Call, they would be confined to the ships for the duration of the time the SLF was in Subic Bay. In addition, we made it clear to them that if they wanted to return to the Philippines sometime in the future to "claim their bride," it would have to be after their tour in South Viet Nam had been completed.

If troops consumed "goof balls," and other kinds of hallucinating drugs, they were brought back to the ships by the Navy Shore Patrol and were held aboard ship. Some of the drugs were so powerful that it would take 36-48 hours before the effects of the narcotics dissipated enough so the person could even sleep. Despite the warnings, there were some troops who suffered through the agony of drug abuse. As I have mentioned before, their logic was "What are they going to do...send me to Viet Nam?"

Probably the most astonishing experience I ever had as an Officer in the Marine Corps occurred while we were at Subic Bay. Col. Mueller directed all Officers of certain rank and above in 2/7 to go to Olongopo and "buy out" a prostitute to take to the Navy Officers' Club at the Subic Bay Naval Base for dinner and dancing. He said to be able to take them to the Base, we would have to pay an unspecified amount of money to the bar where the prostitute worked. He also explained that the amount would be equivalent to the amount the prostitute could make in one night, and that amount could obviously vary. We only had our "jungle utes" to wear to what was considered a formal occasion.

We did what we were told to do. After dinner, Col. Mueller began dancing with his "guest" and he told all of us to join them with our "guests." The Admiral in attendance was the Commander of the Subic Bay Naval Base. When the Admiral and his wife began to dance, Col. Mueller and his guest seemed to intentionally "collide" with the Admiral and his wife. It seemed like a very poor reconstruction of a cheap movie, in my opinion. After the collision, the Admiral and Col. Mueller faced each other, and the music stopped. As they "squared off," everyone else on the dance floor seemed to square off with someone from another branch of the service. The Admiral said, "What do you think you are doing bringing these Philippino whores in here?" Col. Mueller nodded to his guest and said, "She's the best-looking woman in this place." After a few seconds of silence and glaring at each other, the music began, the dancing resumed, and no punches were thrown. For most of us junior officers, that was enough. We retired early and returned our guests to their places of employment.

Special Landing Force Deployment

MY S-4 RESPONSIBILITIES IN 2/7 were not enough to keep me from thinking about the tragic events of the past few months. I missed Sid, Jack, and my troops...every day. During Operations Eager Yankee and Houston, 2/7 had at least two more friendly-fire incidents, one of which was when a "four-deuce" mortar was set in place after darkness. The limited visibility prevented the mortar crew from being able to see possible obstructions. When a fire mission was called in, the first round out of the mortar tube mistakenly hit a tree limb nearby, and the resulting explosion killed the Officer in charge of the unit.

On another occasion, a patrol was sent out to set up an ambush, and for some reason, the Marines did not go out nearly as far as they were supposed to. When the time came to break bus site, the patrol stood up and began to move back toward the position of the unit that sent them out. The patrol had left in complete darkness, but by the time they began to move back to the unit, it had become a very bright moonlit night. Marines on the perimeter of the main position saw them stand up and begin to move toward the perimeter. The Marines in the holes called mortar fire onto the patrol thinking it was an enemy force. Unbelievably, as often happens in those kinds of situations, the first mortar round that struck the ground killed the Radio Operator and destroyed the radio. The shelling of the patrol continued until somehow someone in the patrol was able to fire a red warning flare.

In another tragic event that shocked me personally, the Officer that took over Echo Company from me suffered what appeared to be a permanent, debilitating brain injury after being in command for only about eight weeks. At the time, the reports said he would be in a vegetative state the rest of his life. He was the Company Commander that had been relieved by Col. Love right after I became the CO of Echo Company. (See **"Thou Shalt Not Steal. Thou Shalt Not Kill"** chapter.) He told me previously that he had a scholarship to be a member of a Crew Team at Yale University, or some other Ivy League school, and that he was very anxious to complete his tour in VN. It was related to me that he was moving Echo Company on a search and destroy mission when they stopped to take a break from the heat. He allowed several troops to congregate under a tree with him so they could be in the shade. Enemy forces had placed a mine in the ground under the tree in anticipation of a random chance like that. He had just removed his helmet to wipe his face when someone stepped on the mine and a piece of shrapnel struck him in the head causing the apparent lifelong injury. I was shaken to think that the CO of Echo Company before me had only lasted three days, and that the one following soon after me was also gone relatively quickly. I still wonder about it, but I try not to dwell on it. I thank God daily for His mercy.

Rotation Date

My "Short-timer's Calendar" was nearing completion, and on 7 August, I was in the field with the Battalion Command Headquarters during Operation Houston. Finally, the day had come for me to go back to the world. I left the 2/7 Battalion position by helicopter and arrived at the DaNang Air Base around noon. It was only then that I was told that the flight I was to take to Okinawa was not going to arrive until after dark. I sat there all afternoon trying to make some sense of the last 13 months of my life. The more I thought about it, the turmoil in my head worsened to the point that I could not communicate with those around me who were anxious to get back to the world.

We departed from the DaNang Air Base a little after 2000 hours and arrived at Kadena Air Force Base on Okinawa after midnight. We were transported to Camp Hansen to claim our personal belongings. Within five or six hours, we were airborne and bound for Norton Air Force Base near Los Angeles. However, our plane was running low on fuel and we had to land in Oakland, California to refuel. We were not allowed to leave the plane, and we arrived at Norton AFB in the late afternoon. According to some calculations, it was actually before we had left Viet Nam due to the fact that we had crossed the International Date Line.

Several of us took taxis to the Los Angeles International Airport to catch flights to our home areas. It gave me a chance to see the first of many "ugly Americans" over the next few years. As we stood in line to purchase tickets, a large man dressed in what appeared to be a very expensive suit walked to the front of the line we were in. He slammed his fist down on the counter, and demanded he be served so he could get a ticket on his flight of choice. The airline employees dutifully obliged him while those of us who had been in line for some time waited. There were no bands, no people waving flags, and nobody "thanked us for our service." Rather, the disdain of many people who stood around us was quite evident by their looks, stares, and whispered comments that were spoken loud enough for us to hear some parts of what they said.

After spending several hours in the Stapleton Airport in Denver, Colorado, I arrived in Cheyenne, Wyoming early the next morning. My wife met me at the airport. We traveled to La Grange, Wyoming, where my dad was harvesting wheat. Before going to the field to see if I could help him, I found my mom in our little town to give her a hug. It appeared strange to me that the few other people I saw in the small town that day simply looked at me as if they did not know what to say. Going from Viet Nam to rural Wyoming in approximately 30 hours was a cultural and psychological shock...almost beyond belief.

I was so tired I slept most of the next 48 hours. I went to the field to help my dad harvest wheat, but I fell asleep in the truck. When my family awakened me about dark, we went to my parents' house for supper. I ate very little before falling asleep again. I slept that night and most of the next day and night. I do not know what effect, if any, the sudden isolation at home in Wyoming had on me after leaving South Viet Nam only a couple of days before. Maybe having time to rest and decompress somewhere would have been worse. However, I have always thought that the speed in which I was transported back from the combat zone to the instant deprivation of being without anyone who I had spent the last 13 months with was psychologically damaging. I had no one to talk to about what I was thinking, and I could not adjust to "normal life." I yearned for the month of leave to pass so I could get to Hawaii and hopefully reunite with someone I might have known in the military.

Reporting as Ordered

MY WIFE WENT TO HAWAII ahead of me so she could begin her teaching duties, and I joined her after my 30 days of leave. On the day I was to report for duty, I entered the United States Marine Corps Camp Smith complex on Oahu, Hawaii. Camp Smith was the Headquarters for the Fleet Marine Force, Pacific, or "FMF-PAC", and was the command and control center for all Marine units in the Pacific Theatre, including South Viet Nam. I asked the Gate Guard for directions to the Personnel Office.

I proceeded as directed, and with my Officer Qualification Record, or OQR, under my arm, I entered the office of Major Giovanni who was seated at a desk. It appeared he was engaged in paperwork. I came to attention and stated, "Captain Chamberlain reporting as ordered, Sir." The Major did not acknowledge my presence. I stood at attention for a minute or two, and he finally said, without looking up, "Are you the same Captain Chamberlain that was the Commander of Echo Company, Second Battalion, Seventh Marines?" I acknowledged that I was. He looked up at me and said, "Do you know what a Congressional Investigation is?" I said I did not.

He threw his pen across his desk in my direction, and it landed on the floor. He began to raise his voice and said, "Well, let me tell you." He stated, in a loud voice, that after I buried the Marine, United States Senator Robert Kennedy received letters telling him we were burying Marines in South Viet Nam. As a result, shortly before he was assassinated Senator Kennedy initiated a Congressional Investigation to find out what had happened. He went on to say that someone else had pursued the Investigation after the Senator's death.

In his obnoxious and overbearing comments, he accused me of orchestrating the letters to the Senator. He described the administrative work he had been responsible for as a result of the Investigation, and how I was disloyal as a Marine Officer to get politicians involved in such a volatile issue. His berating of me continued for approximately five minutes while I stood at attention. Finally, he sneered and said that he hoped that I was convinced I had done the right thing. I remained at attention and stood mute. He stared at me, and then he asked me if I had anything to say. I told him I was not aware that any letters had been written to anyone, and especially not a United States Senator. He gave me a look of disbelief and told me to give him my OQR. He then said, "Get out of my office."

Needless to say, it was going to be a difficult situation in Hawaii. I was assigned to Special Services at Camp Smith with no command responsibility. My Duty Assignment was as the Assistant Force Special Services Officer. In a few days, I was told that I would be assigned to be the player/coach of the Hawaii Marine Basketball Team during the upcoming season. Major Giovanni's Office soon discovered that I was due to be discharged on 16 December 1968 because I was a Reserve Officer. As a Reserve Officer I had a three-year obligation to serve after graduation from Officer Candidate School, and then I was to be released from active duty.

A few days later, he called me back to his office and explained that he thought I had augmented and become a Regular Officer when I reported in for duty. As I mentioned before, the term augmented meant that the Marine Corps had decided they wanted me to be a Career Officer. It also meant that they had offered me a regular commission, and that I had accepted it. I assured him that was not the case. He offered me the chance to become a Regular Officer, and I declined. He then said that obviously a mistake had been made at Headquarters Marine Corps in Washington, D.C. He asked me to consider extending past 1 January 1969 to prevent considerable administrative problems because of the mistake. To be helpful, I agreed to extend my tour of duty until 15 June 1969. Little did I know the significance of my decision.

The isolation I felt during that period of time was incredible. I learned a valuable social lesson one day after I called in an order for a pizza. When I arrived to pick up the pizza, the staff at the food establishment refused to wait on me or serve me. Apparently, it was because they could tell I was in the military because of my haircut. Other people I knew said that happened to them, also.

I became the player/coach of the Hawaii Marine Basketball Team, and we played in the military league on the island of Oahu. The military league leader in mid-December was automatically invited to the Rainbow Classic, a prestigious invitational basketball tournament held in Honolulu each year. My team earned that invitation. We lost both games against renowned college teams, but we were competitive, and it was a great honor. That coaching assignment was my one and only time that I was a player/coach throughout my coaching and playing career.

The All-Marine Basketball Tournament was held in Quantico, Virginia, in the late winter of 1969. I was able to get my team to USMC Base Camp Pendleton, California, using space available seats on military aircraft that were occasionally leaving Hickam Air Force Base bound for the mainland. It took three or four days, but the team was finally all there. I had been able to schedule games with several teams in different locations across the country, again flying on military aircraft between major locations. The United States Air Force was very accommodating, and the USAF also furnished military buses when we needed ground transportation. I have always had a very heartfelt appreciation for their assistance.

On our trip across the Continental United States, or CONUS, I was able to take my team to my hometown of La Grange, Wyoming, where we played a team of former college athletes. My players stayed on farms and ranches in the community, and a few were able to ride horses. For most of them, this was the only time they had ever been in rural areas, and they expected the "Wild, Wild West" was going to be like the movies.

We played a game at my Alma Mater, John Brown University, in Siloam Springs, Arkansas. After games in several other cities, we finished our trip with a game against West Virginia State University on our way to the Marine Base, Quantico, Virginia. We finished second in the All-Marine Tournament, and I was selected to be on the All-Marine Team, along with one of my teammates/players Corporal John Funke. It was an honor to represent the Marine Corps in the Inter-Service Tournament at Ft. Hood, Texas which was a basketball playoff between the Army, Navy, Air Force, and the Marine Corps. Corporal Funke and I were awarded All Marine and Inter-Service Sports honors. That year of basketball could be the subject of another book.

As June 1969 approached, I was anxious to be released from active duty to go back to Wyoming. I was due to be released on a Friday. As I walked down a hall of a building in Camp Smith on Monday morning of that week, I was amazed to accidentally see Col. Love. It was a great surprise, and he seemed to be as happy to see me as I was to see him. We talked for a few minutes, and he glanced at the ribbons I was wearing on my uniform. He asked me where my Silver Star ribbon was. I told him I was not aware of what he was talking about. He explained that he had recommended me to receive a Silver Star for my actions during the 1968 Tet Offensive.

I told him briefly how I was regarded at Camp Smith by Major Giovanni and others, and that I had been isolated. He appeared to be angry, and said he was going to look into the situation. He contacted me on Thursday of that week and said he was not going to be able to get me the Silver Star in such a short time frame, but that it was still pending. He also said he was going to have something to give me the next day, my last day of active service in the United States Marine Corps.

When I met him on Friday morning, he pinned two Bronze Star Medals on my chest, one for valor and one for meritorious service. He apologized for the actions against me, or inactions on my behalf, that resulted in me not yet receiving the Silver Star he had recommended. However, he said again that it was still pending. I appreciated his efforts very much, but that was the beginning of my discovering and experiencing what I think is clearly retaliation against me, and others associated with me, as a result of the burial issue. As I mentioned earlier, I discovered Jack's Navy Cross recommendation from me was downgraded to a Silver Star. Other awards I recommended for some of my men were not received at all. While discrimination is often difficult to prove, it does not mean that it does not exist.

I was released from active duty in Hawaii, and I was told to obtain an exit physical from the Veterans Administration (VA) Hospital in Cheyenne, Wyoming. When I arrived there, the physical was nothing more than "feeling my skin to see if I was warm" and "pinching my arm to see if I felt pain." At the conclusion of the five-minute examination, one of the medical personnel unexplainably told me to "never come back." I did not "go back" for approximately 40 years. One of the first medical expenses I paid for myself was for the treatment of a huge infestation of warts on my elbows, knees, and hands, and a cluster wart on a toe, along with large planter's warts in my feet...all of which I had when I was discharged.

As I reiterated previously, after the Colorado American Legion Commander Ralph Bozella insisted that the Veteran's Administration in Wyoming should contact me, I saw an American Legion Advocate who was very helpful. However, an example of the disparaging treatment of Veterans by the VA was when the VA declined to provide a specific medication for me that had been prescribed by a private practitioner for my prostate problem. I had personally paid for the private practitioner. The denial of the medication included the statement, "We do not give medicine that is that expensive to Veterans." The medication that I take includes a warning in the medication narrative that the drug can cause dizziness and "may impair your ability to drive or operate machinery." Those side effects have prevented me from being employed in certain pursuits, and I am unable now to fly an aircraft as a private pilot. As a matter of fact, the drugs prescribed by the VA have prevented me from doing other things that are very important to me for pleasure, relaxation and quality of life. Even though I was exposed to Agent Orange in South Viet Nam numerous times for extended periods, I am repetitively told my various ailments, like my prostate problem and limited skin cancers, do not meet the criteria for any particular consideration from the VA.

While most of my experiences with the Veteran's Administration have not met my expectations, I would be remiss if I did not pay tribute to one VA medical facility that has been excellent to work with. In 2015, I was referred to the VA Hospital in Hot Springs, South Dakota by some other Veterans. Even though it is a three-hour drive to get there, it has been well worth my time and personal expense because I qualified for the Veteran's Choice Program. The service is excellent, and it appears that everyone in the facility is focused on providing the best medical care they can to Veterans. I want to thank them for their service to our nation's Veterans.

Civilian Life and POWs/MIAs

AFTER RETURNING TO CIVILIAN LIFE, I entered Law School at the University of Wyoming (UW) in Laramie, Wyoming. Being confined to classrooms and buildings was psychologically depressing to me. In addition, someone cut the brake lines on my car. It was the assumption of the Campus Police that the reason was because I was a Veteran... and it was 1969. I left UW at the end of the first semester and began a small custom farming operation in the La Grange area. A position to teach and coach basketball and track became available in my high school alma mater in LaGrange in the Spring of 1970, and I was hired to fill the vacancy.

During that time, I was approached by some county political leaders who asked me to consider possibly seeking the office of Goshen County State Representative in the Wyoming Legislature. It was a position that my fraternal grandfather, Earl L. Chamberlain, had held 40 years previously. After winning the 1976 election by a very narrow margin, I began a portion of my life dedicated to public service as an elected official. Interestingly to me, someone who had served with the previous Representative Earl Chamberlain verified that my seat and desk in the House Chambers were in almost the exact same spot where my Grandfather had been seated during his service to our County and State. I was honored to serve 18 years in the Wyoming House of Representatives. My service culminated with me serving as Speaker of the House from 1992 – 1994. In Wyoming, we had a tradition that once a Representative served two years as Speaker of the House, that member did not seek re-election to the House of Representatives, nor did anyone seek to be the Speaker of the House for more than one term. One of the main reasons for that tradition was to prevent the concentration of power in one individual for more than two years.

At the time of the completion of this book, that tradition has been violated and the first Speaker of the House to disregard our tradition has just begun his second term as Speaker. Since the Speaker of the House is elected by his or her colleagues, choosing this Speaker for a second term was obviously the decision of the majority of those elected to the Wyoming House of Representatives. I wish them well.

At about the same time in the 1970s, I was able to re-establish a friendship with another Viet Nam Veteran named Enoch "Bummy" Baumgardner. I had coached two of Bummy's cousins during my one-year tenure of teaching and coaching at Veteran High School the year before I was drafted. Bummy and his wife had visited my wife and me in Hawaii when Bummy met his wife there for R&R. He was a First Lieutenant in the United States Army and served in Viet Nam from September 1968 until August 1969. He was a Mechanized Platoon Leader of a unit of twin 40 mm guns in Battery B, 1st Battalion, 44th Artillery, and was attached to the 3rd Marine Regiment in I Corps during part of his tour. After his return to civilian life in Goshen County, Wyoming, he pursued a degree in Business and began ranching with relatives, which eventually led to him owning his own large ranch southwest of Torrington, Wyoming.

In the late 1970s, Bummy had become involved with the National League of POW/MIA Families, referred to as the League. It was an organization of family members of Viet Nam Veterans and concerned citizens attempting to create a national focus on the issue of the POWs/MIAs that the League, and some other Americans, believed were still alive in Viet Nam. Even though Bummy's participation was on a limited basis, as a member of the League he received information that was distributed regularly by them. Both Bummy and I were very frustrated with what appeared to be a casual effort by the United States Government to attempt to determine whether there were still American POWs/MIAs in Southeast Asia. As a result, we occasionally discussed various things that we could do to possibly aide in the effort. In the fall of 1984, it appeared to us that if evidence were found that POWs/MIAs were alive, a considerable amount of money could possibly be made available to the POW/MIA effort from individual philanthropists and the Ross Perot organization. With that in mind, Bummy and I decided to attempt to make a "low-key," private contact with the Communist Vietnamese Government. We were total amateurs in that kind of international arena, but we believed anything was "worth a try." Our intended approach was to make a cash offer to the Vietnamese for every living American they were holding in captivity. However, we did not want to disclose the idea to anyone in advance because of media exposure and the possible distortion of our efforts.

We decided that the following terms would govern our communication:

(1) Both of us were acting as private citizens;
(2) Our conversations would only involve the issue of possible living POWs/MIAs, not the remains of deceased Americans;
(3) Cash would be offered for each individual POW/MIA who was alive and could be returned to the United States;
(4) Neither of us would claim to have the necessary funds, but we were going to represent the belief that the money could be made available through private sources within the United States; and
(5) The offer would be $1 M for every POW/MIA returned alive.

Keep in mind that we were just two "country boys" trying to make a difference on behalf of our "brothers" who had served our Country in the Viet Nam War. The return of Marine Private First Class Robert Garwood from Viet Nam in early 1979 had inflamed the contentions that American captives were still alive there. PFC Garwood claimed to have actually seen American POWs in Viet Nam before he requested that the Vietnamese government allow him to be repatriated back to the United States. The Vietnamese have always contended that PFC

Garwood originally remained in Viet Nam voluntarily and that he was lying about seeing POWs and MIAs.

Our plan was to try to contact officials in the Vietnamese, Laotian, and Cambodian governments, with our primary interest being the Vietnamese. The focus of any conversations, if we were able to establish any, was to be three basic, initial questions:

1. Are there any American POWs/MIAs in your country?
2. Are there any American POWs/MIAs in a neighboring country?
3. If so, which country?

With that in mind, we attempted to contact the Vietnamese Government. Because the United States Government had no formal diplomatic relations with Viet Nam, the Swiss Embassy in Washington, D.C. was able to arrange telephone contact between us and the Vietnamese Ambassador's Office at the United Nations in New York City. As soon as we made the initial telephone call, the Vietnamese officials seemed to envelop the conversation and future calls with a "cloak of secrecy." When we explained that we wanted to meet with the Ambassador, they asked what the issue was. When we told them we wanted to discuss the POW/MIA issue, the responses from the Vietnamese became even more sinister. We were given code words to use in future conversations, and we were told to make future calls during hours after midnight local time. They also required our full names and social security numbers. After two follow-up calls, a potential meeting was set for 10:00 a.m. on November 28, 1984. The meeting was to be held at the United Nations Building in New York City.

The "red eye" flight we took from Denver, Colorado, was a connecting flight through Atlanta, Georgia. As fate would have it, the flight from Atlanta to New York City was delayed for a few hours because of mechanical problems. Due to the delay, we missed our scheduled meeting with the Vietnamese. When we finally arrived, the U. N. General Assembly Session had already begun. We were told we could send messages to the Delegates from the lobby area, and we attempted to do that. However, the Vietnamese Ambassador sent a message back saying he was occupied and that the meeting would have to be delayed. Fortunately, it gave us a chance to meet with the Ambassadors from Cambodia and Laos prior to meeting with their Vietnamese counterpart.

We sent a message requesting a meeting with the Cambodian Ambassador first. To our surprise, two very young men quickly emerged stylishly dressed in blue suit coats, grey slacks and "penny loafers." We stepped into a side area of the lobby to converse. They introduced themselves as the U. N. Ambassador and Deputy Ambassador from Cambodia. Their response to our first question concerning possible POWs/MIAs in their country was "No." They briefly explained that if any had ever been there, they would have been killed during what is sometimes referred to as the "Killing Fields" era in their country. They also said they were not aware of any Americans being held in captivity in any surrounding country.

We then sent a message to the Laotian Ambassador. A young man appeared who did not seem to have the same level of professionalism that the Cambodians had exhibited. He introduced himself as the Deputy Ambassador. His answer to our first question about Americans being held in his country was negative, but to our intense interest, he said that he was aware of American POWs/MIAs being held in a neighboring country. When we asked him which country, he stated that there were American POWs and MIAs being held in Viet Nam.

Almost as soon as he made that revelation, he was joined by an older man who identified himself as the Laotian Ambassador. When we asked him the same questions we had asked his Deputy, he replied that there were no POWs/MIAs in their country or neighboring countries. We immediately brought the responses of his Deputy to his attention, and he instantly confronted his colleague with an "icy glare." After a few seconds passed, he calmly reiterated his denials. Meeting with the Cambodians and the Laotians first was probably a "blessing in disguise." Thankfully, we had been able to obtain the disclosure of the POW/MIA information from the Laotians before our meeting with the Vietnamese.

The Vietnamese Ambassador finally came out of the General Assembly. As we expected, he appeared to be very professional in mannerism and stature, but his aggressiveness caught us by surprise. As he approached us, we extended our hands out to him to introduce ourselves. Without shaking our hands, he quickly looked at me and, in fluent English, said, "You are Marine Captain Douglas Chamberlain, you were at war in my country in 1967 and 1968, you have three children, you are in the Wyoming Legislature, and you live in La Grange, Wyoming."

Without hesitation, he turned to Bummy and said, "You are First Lieutenant Enoch Baumgardner, you were at war in my country in 1968 and 1969, you have three children, and you live in Veteran, Wyoming." It was an obvious attempt to obtain a psychological advantage, and he was successful, at least initially.

He then asked us what we wanted to speak to him about. With more elaboration than with the Ambassadors of the first two countries, we began to inquire about the possible existence of American POWs/MIAs in Viet Nam. He quickly denied that any Americans were being held in his country "against their will." We explained that, to the contrary, Bummy was part of an organization that had photographs, both from satellite imagery and ground-level photography, which indicated that American POWs/MIAs were in Viet Nam. The claims United States Marine PFC Robert Garwood made when he was repatriated in 1979 were part of the evidence we presented to him. We also assured him that our purpose was not to retrieve the numerous remains of American military personnel that we understood were stored in large warehouses in Viet Nam. We assured him we were only there to communicate with him about American POWs/MIAs that were still alive in his country.

In response, the Ambassador adamantly denied that his country was in the possession of the remains of any American military personnel. In the discussion, we elaborated that we had confidence that we could raise enough U. S. Currency to pay $1,000,000 for every live POW/MIA that was being held in captivity by his government. He replied that it was insulting for us to accuse his country of such acts because they believed that to hold POWs would be inhumane.

When we related to him the conversation that we had with the Laotian Deputy Ambassador earlier, he dismissed the Laotian's comments entirely. As far as the monetary payments were concerned, he assured us that the economy of his country was improving and that they did not need any money from the United States. However, in our preparation for the trip, we had reviewed international economic information and we knew the Vietnamese were in dire need of economic assistance.

The meeting lasted approximately 10 minutes, and it ended with him asking us how long we were going to be in New York City. We told him that we were departing later that evening, and he indicated that he was interested in staying in contact with us. However, because of our lack of any official status, we both believed any further conversation would have to include the support of someone within the United States government.

I am sure it was a clear indication to the Vietnamese Ambassador that we really were amateurs "dabbling" in the real world of international affairs. On the road to the airport, we discussed the possibility that he was indicating to us that he might meet with us again. On the flight back to Denver, we shared the satisfaction that at least we had tried to do something about the POW/MIA issue. We agreed that somehow we should relay our conversations with the three United Nations Ambassadors to someone in the United States Congress.

Soon after our trip, Bummy happened to be watching news coverage of the repatriation of the remains of some United States Military Personnel who had been missing in Viet Nam. He said the Vietnamese Ambassador to the United Nations we had met with in New York City, who told us his country was not in the possession of any remains of American Military Personnel, was standing on the tarmac at the airport in Hanoi as the remains were turned over to United States Government Officials.

A few short months later, I was in Washington, D.C., on a business trip concerning renewable fuels. While I was there, I met with the Wyoming Senior United States Senator Malcolm Wallop. I described the trip Bummy and I had made to the United Nations Building to Senator Wallop and briefed him on the details of the meetings with the three United Nations Ambassadors. I gave him the following reasons that I thought those meetings possibly could lead to the location and return of POWs/MIAs from Viet Nam:

(1) The information accumulated by the League had not been adequately refuted;
(2) Evidence of the existence of POWs/MIAs reported by Staff Reporter Bill Paul on the front page of the Wall Street Journal on Tuesday, December 4, 1984, one week after our meetings in New York City; and,
(3) Despite the denials by the Vietnamese, they had encouraged and arranged the meeting with us in New York, they wanted to know how much money was being offered, and they indicated that they were interested in meeting with us again.

Senator Wallop seemed quite interested, and he suggested that I meet with a new United States Senator from Massachusetts named John Kerry, a Viet Nam Veteran who was on the Senate Foreign Relations Committee. I thought it was an excellent opportunity, so Senator Wallop made the arrangements. I was told that Senator Kerry was in a Banking Committee Meeting that day, but due to time constraints, I could meet with Senator Kerry in the coat room adjacent to the Committee Room during a break. I was directed to the coat room, and Senator Kerry entered shortly thereafter.

I stood, introduced myself to him, and extended my hand. Without acknowledging me, he just stared at me in silence. Senator Kerry appeared aloof, disinterested, and he had an insolent air about him. I described very briefly my military experience in Viet Nam and my frustration with the POW/MIA issue, especially with the apparent lack of effort by the United States Government to pursue possible sightings of American Military personnel who may still be alive in Viet Nam. I also quickly described the meetings Bummy and I had with the three Ambassadors at the United Nations Building several months previously. He stood mute the entire time, and when I paused to allow him to interact with me, he simply said, "Is that all?" When I acknowledged, "I guess it is," he mumbled a "Thank you" as he turned and left the room without any comment.

Senator Kerry's attitude and demeanor that day, and the obvious disdain he displayed toward me as a fellow Viet Nam Veteran, appeared to be that of the majority of the people in our Country for decades. It also convinced me that there was a conscious effort within the United States Government to deny the possible further existence of any POWs/MIAs.

Looking back at the efforts Bummy and I expended concerning the POW/MIA situation, I can now see more clearly what some of my motivations were. I vividly remember how the actual daily trauma and acute feelings of guilt about the burial issue were almost totally consuming me. The sleep deprivation and the need to be "continuously involved in something" I experienced then have continued for the decades since. I seldom have been able to accept criticism from people that have no idea what is actually going on in the recesses of my mind... but I have never been able to bring myself to attempt to explain to anyone personally what I have finally disclosed in this book. The continuous mental agonizing has slowly helped me identify the events or issues that are the most "explosive" for me and which ones are my "triggers." I avoid them at all costs, but it has led me to a life of significant personal isolation. My personal mental trauma has prevented many of the most important people in my life from being able to understand me, and why I have difficulty developing and maintaining personal relationships. However, as strange as it may seem, my responsibilities to my constituents in the public service portion of my life appear to have given me focus and direction as I have emotionally "careened" through life.

The Search for Truth

THE "STRAW THAT BROKE THE CAMEL'S BACK," and eventually led to the writing of this book, was because I had been dealing with the burial issue for nearly 50 years, and I wanted to see my OQR. The answer from the USMC at the time I started writing this book was that my military records were still missing. An inquiry from Wyoming United States Senator Mike Enzi in 2017 produced copies of part of my records for the first 18 months of my service without any elaboration or explanation for the previous claims from the Marine Corps that my records were missing. In the partial disclosure, there were very few records from my service in South Viet Nam except information indicating that I was recommended to receive a Silver Star for Valor... that I have never received. The partial disclosure made me even more determined to obtain my OQR. I had anticipation, as well as a foreboding, that my missing records would contain evidence that would validate my belief that the burial issue included a cover-up, and that I was the person accused of possibly violating the Uniform Code of Military Justice (UCMJ).

During these many years, I have agonized continuously, and I have been consumed with questions surrounding the burial of the Marine. Who gave me that order? How many people were involved in the cover-up? What action, if any, was taken against those involved? What happened to the remains of the Marine we delivered to the morgue at the DaNang Air Base? Were the remains delivered to the family with appropriate military honors? If so, what was the family told about the entire situation? Where are the military records concerning this entire issue? Where are my military records? What were the results of the Congressional Investigation initiated by Senator Kennedy, if there actually was one? What action was taken, if any, as a result of the investigation?

It would appear to me that the cover-up must have prevented any accurate, detailed, definitive resolution of the issue because I was never contacted by anyone concerning what actually happened, except Col. Love after I initially reported it to Col. Mueller. What was the actual basis for the "offer of recognition" made to me that supposedly came from the Commanding General's Office of the 1st Marine Division? (See **The Rest of the Story** chapter.) I had almost abandoned my pursuit of the answers to these many questions, and the truth seemed to be disappearing into the sunset of my life. Then... as if by some miraculous revelation, the facts relating to the many questions above began to be revealed and verified in a manner, and at a speed, I never would have believed was possible. The following facts and findings defied my wildest imaginations.

The Truth, the Whole Truth, and Nothing but the Truth

BEFORE I ELABORATE on the details of truth that is to follow, I want to introduce and pay tribute to the people who have brought the facts forward that have validated my memory of what happened in 1968 in the Viet Nam War. This book would not have been brought to the finality that it encompasses without the professional work of Paul Semones, P.E., the owner of Semones Forensic Engineering located in Pleasant Grove, Arkansas. He graduated from John Brown University in 1999. His immense knowledge of on-line research and professional skill in the understanding of the human psyche is unmatched by anyone I have ever known in my life. I will always marvel at his organizational skills and his assistance in bringing together the mass of information in the chronological manner needed to shed intense light on something that happened one-half century ago.

Mr. Semones was able to produce investigative findings that authenticated the history of events that occurred during 1968 in South Viet Nam, the proof of which had lain dormant for five decades. Astonishingly, he accomplished nearly all of this in a matter of two weeks. As this information reveals, it is nearly impossible to quantify and qualify adequately what he has done for the writing of this book. Paul Semones was able to locate additional information as a result of the research performed by Janet Foster, Manager-Constituent Services in Arkansas United States Representative Steve Womack's local office in Rogers, Arkansas.

Representative Womack is a friend of John Brown, III, who was the President of John Brown University (JBU). The Womack Representative

District includes Siloam Springs, Arkansas, where John Brown University is located. Representative Womack is a fellow Veteran, and I will be eternally grateful to both he and Ms. Foster for their assistance.

I have also received immense assistance from Wyoming United States Senator Mike Enzi, and his State Director Dianne Kirkbride. Mike has been my personal friend for over 40 years. Senator Enzi is an outstanding public servant and supporter of military Veterans.

In addition, Wyoming United States Representative Liz Cheney has been exceptionally helpful and supportive. Representative Cheney is a cornerstone of the political foundation of the Wyoming Republican Party. She is undoubtedly one of the most influential and effective Freshmen Congresswomen ever to be elected to the United States House of Representatives. Representative Cheney selected me and the record of my military service in South Viet Nam to be part of the Veteran's History Project that will be filed in the historical vaults of the United States Library of Congress in Washington, D.C. for review by future generations of Americans.

With this cast of outstanding Americans, I was able to renew my search for definitive truth about my burial of the young Marine in South Viet Nam. It took me 50 years, almost to the month, to discover what the inclusive details actually were. To reiterate, I had nearly exhausted my resources of information as I neared what I thought was going to be the completion of my book/memoirs, such as it was, in mid-2017.

However, in the fall of 2017, John Brown, III had become aware through casual conversations with me that I was engaged in a serious effort to write a book encapsulating the memoirs of my military service in South Viet Nam during the Viet Nam War. He inquired about my progress occasionally, and I finally told him I thought I had located all of the material I could to verify my writings. I also voiced my disappointment about having so many of the questions that had plagued me for so many years go unanswered.

In one of our conversations in late 2017, he mentioned that he was having a Forensic Investigator do some research for JBU that included biographical details of all JBU students who were killed in World War II. It was an effort to not only memorialize them, but to honor their efforts and the sacrifices of their families for posterity sake. John suggested that it might be of help for me to allow this Investigator, Paul Semones, to read my manuscript.

I did not give it much thought until he mentioned it again in early February of 2018, and he again urged me to at least have Forensic Investigator Semones look at the chapter of my manuscript entitled "**Lives Changed Forever**." Because of my utmost respect for John as a person, as well as his outstanding judgement, I decided to allow Paul Semones to review part of my manuscript dealing with the burial. I had never met Mr. Semones personally. However, the decision to engage the efforts of Paul Semones was one of the most important decisions I have ever made that *"changed my life forever"* from a positive perspective.

John Brown, III, gave Investigator Semones the limited portion of my manuscript to review that dealt with the burial of the Marine. Mr. Semones called me a few days later and we discussed my efforts generally, and the purpose I had in writing what is basically my Viet Nam War Memoirs. When we discussed the burial issue, I sensed a certain amount of ambivalence in his voice. As we pursued that issue, I also sensed that he did not really want to explore an allegation that happened one-half century ago unless there was strong evidence to verify the legitimacy of a story that arguably could have seemed somewhat preposterous at that point in time. However, out of his respect for John Brown, III, I think, he agreed to give the matter at least a cursory glance.

Paul has demonstrated that he is a consummate professional with the ability of a genius in the field of internet forensic research. He initiated his efforts in late April 2018, and within 14 days he had authenticated many of my contentions to the point that he apparently began to think the details of the burial were irrefutable. My allegations concerning the burial of the Marine are in the previous chapters of this book. The following is a summary of what the Semones research revealed.

Paul surmised that the best way to discover the entire story about the alleged burial was to attempt to personally identify the Marine my men and I had found lying in the bomb crater. To do that, he had to narrow the leads down to a handful of possibilities. Even though many Marines had died in that area during Operation Worth and in the months preceding Operation Worth, Paul could not locate any KIA records that seemed to match the information that I had given him concerning the approximate location, manner of death, or the assumed timeframe I was alleging. He almost abandoned that particular avenue of research, but before he did, he performed some generic searches on the names of some of the higher-ranking officers in the various Battalions of the 7th Marine Regiment.

His research revealed an article published in the *Leatherneck Magazine* in August 2010 entitled "The Doom Patrol" written by Marine Lieutenant Colonel Jack Wells (Retired). The substance of the article seemed strikingly similar to the burial story in my book.

He contacted Lt. Col. Wells and was able to confirm that the Wells story was actually about the same Marine. However, Lt. Col. Wells had written it from the perspective of the Marine unit that had actually abandoned the Marine and supposedly recovered his remains sometime later. Lt. Col. Wells revealed to Paul that the name of the dead Marine I had spent my life shamefully pondering was Private First Class Michael J. Kelly from Findlay, Ohio.

Without that information, the mystery I had contemplated for so many years would have been very difficult to solve. After Paul's initial discussion with Lt. Col. Wells, Paul and I discussed whether we should record interviews that we might be able to arrange in the future. Instead, we decided that we would attempt to have both of us present telephonically when either of us spoke to persons of interest about the subject.

The information in the Wells "Doom Patrol" story came from an embellished Marine Corps Public Relations Release written in 1968 that was intended to make that Marine Patrol look like heroes. Even with that said, their efforts were definitely laudatory, in my opinion. However, Lt.

Col. Wells told us that he knew that his story was only partially true in that the Doom Patrol had only recovered the leg of the deceased Marine, not the entire remains of PFC Kelly, as was implied in his story. Lt. Col. Wells also admitted to Paul and me that his intentions were a good faith effort to prevent the family of the abandoned Marine from having to endure the bitter truth about their loved one who had given his life for his Country.

Armed with the identification of the deceased Marine, Paul was able to establish contact with PFC Kelly's son. In order to be able to examine the records of PFC Kelly, we needed his son's permission. Paul was able to obtain that. United States Representative Steve Womack, who represents the Third Congressional District in Arkansas, was able to then have his staff person, Janet Foster, retrieve PFC Kelly's records from the National Personnel Records Center in St. Louis, Missouri.

The information in PFC Kelly's file enabled Paul and me to confirm all of the essential details of this sordid tragedy. The attempts to cover up the burial information by those in powerful positions in the military bureaucracy at that time are "bone chilling" and disheartening. Paul was able to establish that persons included in communications concerning PFC Kelly's death, burial, and the recovery of his remains by me and the volunteers from my Company, included the President of the United States in 1968, Lyndon B. Johnson.

Paul established the area in the I Corps region of South Viet Nam where PFC Kelly was killed by the use of map coordinates. PFC Kelly was a member of Charlie Company, First Battalion, Seventh Marines and on 15 February 1968, the day of his death, he was the point rifleman of the Platoon that was in the front of Charlie Company during Operation Pursuit. The Operation was one of several in that area during the months following the 1968 Tet Offensive. Those Operations were designed to prevent renewed ingress of North Vietnamese Army units from Laos eastward through the Happy Valley region into the greater DaNang area.

Military Records indicate that Charlie Company was ambushed by enemy forces, and that PFC Kelly was killed instantly. The records also disclose that he had devastating injuries to his left leg, indicating that the leg was severed by an explosion of something, possibly a hand grenade. Several other Marines were killed and wounded during the fighting that ensued.

The Company Commander of Charlie Company was just returning from R&R, and he had not yet resumed command of the Company after his brief absence. The Acting Company Commander called for air and artillery support and, when he believed he had proper tactical control of the situation, he called for choppers to medevac his casualties. When the aircraft arrived, they encountered intense enemy ground fire as they approached the makeshift landing area dangerously close to where the Company was engaged with the enemy forces. The CH-34 Korean War era choppers did not have enough power to lift out all the casualties and, for unknown reasons, PFC Kelly's remains were the only ones left with Charlie Company. The tactical situation was such that the Acting Company Commander decided that he should consolidate his Company with the rest of the 1/7 Battalion that was behind them.

As they began to consolidate forces, carrying the remains of PFC Kelly was apparently burdensome, and the Acting Company Commander notified those in command of 1/7 that he believed he was jeopardizing the safety of his entire Company by continuing to carry the dead Marine. He sought permission to leave PFC Kelly behind with the understanding that he would be retrieved later. The permission was granted, and his remains, including his left leg, were left lying in a bomb crater.

In a conversation with the Combat Photographer who was with Charlie Company at that time, the Photographer told Paul and me that he carried the entire remains of PFC Kelly away from the ambush site by himself. He said he finally collapsed from exhaustion. The Photographer also said that when he regained consciousness, the remains of PFC Kelly were lying in a bomb crater close to where he had collapsed. He said Charlie Company then continued the withdrawal to consolidate with the rest of 1/7. Upon the Company's return, the 1/7 Battalion Commander was notified that a Marine had been left behind.

Further interviews Paul and I conducted, including military records, both indicate that the CO of 1/7, some members of his Battalion Staff, and the Charlie Company Commander who had just returned from R&R decided to solicit volunteers, to return to the area where PFC Kelly had been left, to retrieve his remains. Eight young Marines, under the leadership of a Sergeant, volunteered for what could reasonably have been considered a possible "suicide mission." Their effort was later referred to as "The Doom Patrol" in the Wells article mentioned previously. They were given air support to cover them, parts of which were Marine F-4 Phantom jets. The Marine volunteers left the 1/7 base under the cover of darkness, and they were able to reach the bomb crater where the body had been abandoned shortly after daylight. North Vietnamese units engaged them, so they took the severed leg of PFC Kelly and returned to the 1/7 location.

After they left the rest of the remains in the bomb crater, the Doom Patrol notified the FAC for the F-4's flying cover for them that the NVA units were around that bomb crater and they told the FAC to have the F-4s "bomb the bomb crater." The pilot of one of those aircraft was the one who I had the "chance" conversation with about Distinguished Flying Crosses during the flight back from Hawaii when I went on R&R a couple of months later. (See **R&R** chapter.)

Records indicate that the Doom Patrol returned the leg to 1/7 where they were hailed as heroes by the 1/7 CO. He related information about the recovery effort to the media at the time and made the patrol members available for interviews. **This same 1/7 CO later wrote a book in which he lauded the efforts of the Doom Patrol and allegedly stated that their efforts were more heroic than the efforts of Charlie Company, 1/7 at the Chosin Reservoir battle during the Korean War.**

To keep the efforts of the Doom Patrol in their proper perspective, the Marines who fought at the Chosin Reservoir in the deadly cold of the Korean winter suffered hundreds of casualties, both killed and wounded, and the rescue and evacuation of the 1/7 Marines at the Chosin Reservoir is recognized as one of the major feats in

United States Marine Corps History. Some individuals we interviewed said they understood that the members of the Doom Patrol had been nominated to receive bronze stars for valor, or bravery, by the 1/7 CO.

One important fact that the 1/7 CO chose not to relate to the potential readers of his book, was that the Marines under his command did not return all of the remains of PFC Kelly. The sordid truth is that they left all his remains behind except for his left leg. He also avoided the admission that the integrity of the 1/7 CO became compromised at that time. If this entire burial issue is, among other things, one of integrity, and in my opinion it is, Marine General Charles C. Krulak's remarks entitled "**DEFINITION OF INTEGRITY**" that he delivered to the Joint Services Conference on Professional Ethics 27 January, 2000 is pertinent. The following are very poignant words near the conclusion of his remarks:

> "...discern between what is right and wrong; act on what you have discerned to be right, even at personal cost; and influence others to do the right thing. And always, always, remember that no one can take your integrity from you...you and only you can give it away."

We could not find records that he ever indicated any concerns about the ensuing cover-up of the tragedy by Marines in his Battalion, the effects of which seriously damaged many other Marines psychologically forever. The extensive attempts to cover up this disgraceful and dishonorable act can only be considered as a tragic violation of the foundational principles of United States Marine Corps tradition, honor, and loyalty, in my opinion. I believe the shocking facts are of historic proportions.

Military records indicate that Marine protocol was followed as far as the initial notification to the Family of the death of their fallen Marine. However, my understanding of the PFC Kelly records as we studied them is that **the Family was unaware that the remains returned to them** were not their son's entire remains until just a few hours before the funeral service. None of the official records we located indicate that personnel at higher levels of Marine Corps Command in the Viet Nam Theatre of Operations had any knowledge of the efforts to conceal any facts from the Gold Star Family at that particular time. However, after an apparent Family letter of inquiry to PFC Kelly's unit, his Platoon Commander wrote a letter of condolence to the Family on 2 March 1968 saying that his entire body had been recovered. Whether he wrote in error or was intentionally misleading the Kelly family is unclear to me. The Officer stated:

> "We lost two more men but got his body out shortly afterwards. We removed it to a spot where we could get him out the next morning by helicopter. The next morning we tried to get him out and we were attacked by the VC who pushed our platoon down off the hill. We sent the 8-man squad up to get him that night and his body was safe within our lines within 18 hours after we had left it."

The truth was that only his leg "was safe within our lines..." That Marine Officer was also killed just a few days after writing the letter. When the initial remains of PFC Kelly were finally returned to Findlay, Ohio, for burial in March, a Family member, who was a Korean War Veteran, was asked by the parents of PFC Kelly to view the remains prior to a closed casket graveside service. We did not see any evidence that the Korean War Veteran indicated that the remains consisted only of a leg. **However, military records indicate that the remains were "unviewable."**

Tragically, when the Kelly Family buried what they thought were the entire remains of their son on 28 March, 1968, the Department of the Navy already knew that my Company had, in all probability, discovered PFC Kelly's partially decomposed body and delivered it to the morgue at the DaNang Air Base. The responsible officials chose not to notify the family of our discovery before the funeral. The records indicate that officials made the decision so they would be able to verify that the remains we had delivered to the morgue in DaNang, including the "dog tags" bearing the name of PFC Michael J. Kelly, were actually his.

Our discovery of PFC Kelly's remains set off a "tidal wave" of cover-up, denial, and retribution. Understandingly, heartache and mental trauma concerning PFC Kelly's burial may still exist in the minds and souls of PFC Kelly's son and his son's mother, as well as the Kelly family members that are still alive. I can assure you that the issue is still a vivid memory in the minds of many of the surviving Marines who were involved, including me.

I want to make it absolutely clear that the disclosure of these verified records and interviews concerning PFC Kelly should not in any way diminish the bravery of those volunteer Marines in Charlie Company, 1/7 who risked their lives to recover his remains. However, the motives for the embellishment of the Doom Patrol actions by others are more than mere conjecture and speculation by me.

The entire truth concerning what remains were recovered and what were left behind, the supposed reasons why, and the cover-up of the truth concerning why the remains of PFC Kelly were left behind will be for the readers of this segment of Marine Corps history to contemplate. Similarly, you, the readers, will ultimately need to formulate your own opinion of the Marine legacy those involved in the PFC Kelly saga have left behind.

With due respect for the family of PFC Kelly and my Marines in Echo Company, Second Battalion, Seventh Marines, the following facts disclose the irrefutable truth the Semones Research has established. **As the late Paul Harvey would have said, "This is the rest of the story."**

In the original chapter named **"The Rest of the Story,"** I reported the order I received to bury the remains of a Marine to the Commanding Officer of 2/7, which was my organizational Battalion and my Battalion Commander. **My disclosure led to an apparent Marine Corps Article 31 investigation led by Col. Love that I was not aware of at the time.**

Also unbeknownst to me, one of my Marines, Corporal Willie Williams, wrote a letter to President Johnson on 25 March 1968 describing the burial of a Marine by me in South Viet Nam. **(Attachment A p. 304)** It is not known how the letter reached the President so quickly, but it was received in the White House Mail Room at 2359 hours on 29 March, 1968, two days after my volunteers and I had exhumed the remains and delivered them to DaNang. By the time President Johnson received the letter, the first of what became two funerals for PFC Kelly had already been conducted by the Kelly family. Records also indicate that the Williams' letter was forwarded to the United States Department of Defense where it was received on 8 April, ten days later.

It is unclear how United States Senator Robert Kennedy became involved, as was alleged in the accusations made against me by the Personnel Officer when I reported for duty to United States Marine Corps Camp Smith, Hawaii in the fall of 1968. It is possible that it was a result of the Kelly family seeking Congressional assistance as described in a Casualty Assistance Call Report dated 2 April 1968. In the narrative of that Report, it stated that the Kelly family had been notified on 19 February 1968, that their son had been killed on 15 February, and burial arrangements had been made pending the return of PFC Kelly's remains.

After waiting several weeks, the Kelly Family had apparently, and understandingly, become concerned that their son's remains had not been returned to them. (Attachment B p. 309) If the letter they had received from PFC Kelly's Platoon Leader alleging that his entire remains were recovered intact was correct, they could not understand the delay and what had happened. (Attachment C p. 312 – exact reproduced copy of the handwritten correspondence is also included for clarity to the reader) What was represented to the Kelly family to be the complete remains of PFC Kelly were not made available to the Family until approximately six weeks after his death. The Kelly family was told that it was due to identification difficulties.

It had to have been impossible for United States Military Officials at the morgue in DaNang to positively identify whose leg it was. Apparently, the Officials relied upon some sort of eye-witness accounts or other speculative information. The painful truth we now know is that only his leg was returned to the family initially and the cover-up that began shortly after PFC Kelly's death was continued by the growing number of Marine Corps and Navy personnel involved.

The cover-up apparently included the Certificate of Death which states that the cause of PFC Kelly's death was "DUE TO GUN SHOT WOUND ENTIRE BODY." However, no authenticated medical evidence could have established that fact because the only physical evidence that initially existed was a portion of someone's leg. The possibility exists that the cause of death was substantiated by eye-witness accounts, but if that evidence existed, we did not discover it. **Regardless, the continuing effort to perpetuate the cover-up was leaving a more encompassing trail of deception, and the Kelly family was being tragically victimized in the process.** As I have stated previously, the deceitfulness of those who were involved in this cover-up has created life-long psychological trauma for me and some members of my Company.

A letter dated 10 April 1968, sent to Corporal Williams from the Assistant Secretary of the Navy stated that President Johnson had learned that the Commandant of the Marine Corps had already been notified of "similar allegations" and that an appropriate Inquiry had been initiated. (Attachment D p. 318)

The apparent result of the Inquiry was a Memorandum dated 14 May 1968, from the Deputy Director of Personnel to the Commandant. In that document, the Commandant was told that only a portion of PFC Kelly's remains were initially recoverable. **It then states that my Company discovered the "unrecoverable" remains, and that the Tactical Commander in the field ordered me to bury the remains because he thought it would not have been advisable to send the rest of the** remains to the Family since a funeral service had already been held. (Attachment E p. 319)

We still have not been able to determine who the Tactical Commander was that gave me the burial order, and how he supposedly knew when the funeral was held for PFC Kelly. **The Tactical Commander's statement to the Article 31 Investigators that a funeral had been held was false. It is also evidence, in my opinion, that the Tactical Commander knew the details of the original false statements about the efforts of the Doom Patrol.**

The horrific facts were that the funeral had still not yet been held at the time he gave me the burial order, because the United States Navy medical authorities had not been able to identify the leg returned by the Doom Patrol. The cover-up was enhanced by alleging in the Investigation Report that I was told to bury the remains in a known location so "they could be recovered later, if necessary." It does not take a "rocket scientist" to know what the meaning of the outrageous words "if necessary" was. Obviously, those two words could only have meant that those in command of 1/7 at that time would have had yet another way to perpetuate an additional lie about the burial order if the previous lies they had already created were discovered in the future.

It is also an entirely false statement that I was told to bury the remains in a "known location." The mandate to me was, "Do as you are told. Don't rock the boat. This is an order. Bury him." It was an absolute order to cover up the truth and to participate in the deception concerning a Marine whose unit had abandoned him. An additional insult to me personally is the complete lie that, "The remains were buried on 25 March 1968, fifty meters from the Company Command Post." There was no Company Command Post. We were in the jungle and, as I stated previously, I buried him in the bomb crater where we found him after I had the confidence that I could see the location from the air.

The deceit had resulted in a tragic delay of almost six weeks before the actual partial remains of their loved one was returned to the Kelly family. In addition, the military representatives perpetuated the unconscionable lie that it was his entire remains until shortly before the service on the day of his first funeral. As I was taught in my youth, "One lie usually leads to another."

In a continuation of the cover-up, a letter from the Head, Casualty Section of the United States Marine Corps to the Kelly family explained that a patrol had been "inserted" to recover their son's remains, as if by some plan. They chose to tell that lie rather than to disclose that we discovered PFC Kelly's remains after they had been abandoned by members of Charlie Company 1/7, and that they were left lying on the ground in the jungle for well over a month.

More lies were perpetuated with the following statement: "I hope, however, you will understand that earlier complete recovery of your son's remains was not possible due to the tactical situation and the inaccessibility of the location." (Attachment F p. 321) That inept, false statement is shocking to me as a Marine Officer.

In a letter to Senator Robert Kennedy dated 2 May 1968, the Marine Lieutenant General who authored the document said that his correspondence was "in further reply to your inquiry of 2 April 1968 regarding Corporal Willie Williams...and your recent letter regarding Private First Class William P. Kelly...U.S. Marine Corps." That letter did not have the correct service number for PFC Michael J. Kelly in the text of the letter, and it had to be corrected with a handwritten note later.

By coincidence, there was more than one Marine PFC with the last name Kelly in the Marine Corps at that time who were both in the Viet Nam Theatre of Operations. Quite obviously, Senator Kennedy had received communications from someone, and the Senator had contacted the Marine Corps previously with some questions.

The Lieutenant General involved in the communication to Senator Kennedy was the Marine Corps Assistant Director of Personnel, and he appears to have continued the perpetuation of deceit. In the letter to the Senator, the admission was made that the Tactical Commander that gave me the order to bury PFC Kelly's remains was aware that the Kelly family had been deceived, and that only partial remains had actually been returned to the Family initially. The desperate cover-up included the statement that members of my Company were made aware that "superior authorities within the 1st Marine Division" had countermanded the order given to me to bury PFC Kelly's remains and that an order had been issued to have the remains exhumed.

The truth was I was furnished the aircraft resources I requested to go on the volunteer mission with my men to recover the remains. I was not ordered to do so. The dishonesty continued when the Marine Officer stated that the Kelly family knew for some time that the initial remains were only partial. Beyond my reasonable and logical comprehension, the text of the letter contains what appears to be a veiled warning: "The release of information of this nature is normally subject to rigid control. It is furnished to you as a member of Congress knowing that it will be subject to discretion as to its use and dissemination." (Attachment G p. 322 – exact reproduced copy of the correspondence is also included for clarity to the reader.)

Was the truth about a Marine being abandoned and left behind in Viet Nam supposed to be subject to discretion? Tragically, Senator Kennedy was assassinated approximately one month after the date of that letter. We did not discover what letters, if any, had been initiated by Senator Kennedy to the White House, the Department of Defense, the Department of the Navy, or the Commandant of the Marine Corps that may have led to the response in Attachment G.

As further evidence that the cover-up was carried out, the Memorandum to the Commandant of the Marine Corps dated 14 May 1968, concerning the issue factually does not completely parallel the letter sent to Senator Kennedy. We did not find any evidence that issues surrounding PFC Kelly were pursued by anyone in Congress after the Senator's tragic death. However, the facts that the Kelly family did not receive their son's remains until approximately six weeks after PFC Kelly's death, and that the remains only consisted of an unidentifiable leg, were omitted.

The Casualty Assistance Call Report dated 2 April 1968 confirms that the Kelly family had sought "Congressional assistance" after receiving the letter from PFC Kelly's Platoon Commander stating his body was recovered intact followed by the ensuing delay. We did not find a record stating from whom the Congressional assistance was sought or what action was taken, if any.

Paul and I discovered two other pieces of information that were of interest to us. One was a rough draft of a letter that was supposedly going to be sent to PFC Kelly's parents from the Marine Officer who was the Inspector-Instructor at the Marine Corps installation in Lima, Ohio. In the draft, an offer was made to the Family to have their son's remains buried at sea instead of being returned to them. It also offered to have his remains cremated prior to the return to them. The offers concerned the first remains consisting of the leg. Was that standard military policy…or could it have been part of the cover-up so the remains could not have been viewed? We could not verify if the Kelly family ever received that letter. If they did, they obviously declined the offer.

The second issue was a RUMHLA Unclassified Message number 2284 dated **19 February 1968**. As I mentioned previously, this document is sometimes referred to as a **Death Report. It contains the personal data concerning the death of PFC Kelly. In paragraph "D." it says, "GM UNVI WABLE PORTION OF REMAINS RECOVERED." In paragraph "I." it says, "INVES RPT WILL NOT BE SUB." Does that mean that there was an "unviewable" portion of remains that were recovered and an investigation report concerning the remains that was not going to be revealed to the Family?** If so, why? These are some of the mysteries surrounding the death of United States Marine Michael J. Kelly that still remain unsolved.

In mid-July 2018, I was engaged in a telephone call with Paul Semones about my manuscript. I had told him previously about my discovery of two 8 x 10" pictures in a scrapbook I had recently found. I had just disclosed to him that I was not sure who all of the people in the pictures were when, for some reason, I turned the pictures over. On the back was yet another astonishing disclosure. One picture was of four Marine officers, one unknown Marine, and one civilian. The civilian was shaking hands with Sid. **(Attachment H p. 326)** On the back is writing that identifies the Officers as: General Robertson – CO of 1st Marine Division; Lt. Col. Mueller – CO of 2nd Battalion, 7th Marines; Sid; and me. **The civilian is identified as Joe Alsop – Syndicated Columnist – Washington, D.C.** The notation also says: "Arriving for briefing at Echo Company Combat Base on Hill 190." The picture is date stamped **MAR 30 1968. (Attachment I p. 327)**

The other picture is of those of us in the first picture, as well as Jack carrying a rifle, and additional Marine personnel walking up Hill 190. **(Attachment J p. 328)** The notation on the back of that picture says: "Moving to top of Hill 190 for briefing of tactical situation in E-2/7 T.A.O.R. and <u>E-Co. appraisal of Operation Worth</u>." **(Attachment K p. 329)** I do not remember what the stated purpose of the briefing was, and why Sid and I, an obscure Company Commander and my Executive Officer, were asked to brief the Commanding General of the 1st Marine Division and a nationally known member of the media...**only three days after I had led my volunteers to retrieve the body of a Marine who had been left behind by 1/7...and who 1/7 personnel had ordered me to bury.**

Paul ran a quick search while we talked and he revealed to me that Joe Alsop had been, and possibly still was at that time, a <u>United States Central Intelligence Agency Agent</u>. The records Paul found indicated Mr. Alsop had a colorful career in Asia that began before World War II, both as a Journalist and in a more active military role with the Flying Tigers in China. He was taken into custody by the Japanese in China on December 7, 1941, but he was able to be released six months later by claiming his status as a Journalist. By the 1950s, he was using his Journalism Career as a cover to gather intelligence for the C.I.A. in the Philippines and elsewhere. I am absolutely certain Sid and I did not mention the burial issue during that briefing, but now I am totally mystified as to why a General and a Journalist/probable C.I.A. Agent came to my remote location for a briefing. Was this part of the cover-up and to see if we would mention the burial issue, or was it a "coincidence?" Be that as it may, these pictures and the events surrounding them may be a subject of possible investigation by Semones Forensics, if Paul has an interest in pursuing it in the future.

The Kelly family experienced trauma generated by this issue beyond anything I can imagine. The crushing pain of the second funeral they had to endure is unfathomable. They exhumed PFC Kelly's casket and had his additional remains my Company recovered interred with his partial remains. My heart aches for them. What I have suffered through the years concerning PFC Kelly pales in comparison to what can only be imagined was their devastating agony. At the time of the writing of this book, the parents of PFC Kelly are both deceased. **However, I trust that the culmination of our investigation into PFC Kelly's death, the handling of his remains, and the disclosure of the truth surrounding it will be of some solace to his son, his son's mother, and to my men in Echo Company, 2/7.**

Semper Fi

SEMPER FIDELIS IS THE TIME HONORED "motto" of the United States Marine Corps. It is a Latin phrase that means "Always Faithful." United States Marines commit to honor their Marine brothers in the spirit of that sacred pledge and, at the time of the completion of this book, Marines have almost always done so for the 243 years since the Marine Corps was created as a military arm of the United States of America, and a component of the United States Navy.

In the case of PFC Michael J. Kelly, a few Marines and civilians were contemptuous in their deceit, lies, and cover-up concerning the mortal sin of leaving another Marine behind. In this book, I have tried to acknowledge the necessity of a tactical decision that could possibly lead to the abandonment of a fellow warrior for a limited period of time. However, in my opinion, only God can forgive those who would intentionally leave a Marine behind...alone on the ground for well over a month...to then be discovered by other Marines...and then continuously lie about their deeds to conceal their betrayal of our sacred honor. The extent of the dishonesty by those in positions of military command that allowed this unthinkable, extended chain of events to happen, and the continued actions of both military and civilian personnel who perpetuated and were complicit in the following cover-up shocks me morally, intellectually, and as a patriot of our great Country.

The entire issue still makes me nauseous. I have made every attempt to keep my anger from overcoming my better judgement for these 50-plus years since we found PFC Kelly. I hope solving this horrible mystery will restore some of the lost confidence and pride to my men who served under my command in Echo Company, Second Battalion, Seventh Marine

Regiment, First Marine Division who are still alive, and that it will calm their spirits and bring peace and comfort to their souls. I also pray that this will bring to them the realization that, as their Commander, I would never have abandoned this tragic issue until I was laid in my final resting place.

This book would not be complete without the final consideration of PFC Kelly's family. On November 17, 2018, I traveled to Findlay, Ohio, to meet the family members who were still alive and who would agree to meet with me. I was accompanied by four outstanding Americans who believe in honor, country, respect, and loyalty. Those who joined me were, in the back row (p. 264) from left to right, former **President of John Brown University John Brown, III**, who traveled from Siloam Springs, Arkansas; former **Northwestern University Psychology Professor Dr. Donald Catherall**, who drove to the occasion from his home in Chicago; me; and the **USMC Commandant's Liaison to the United States Senate Colonel Jon M. Lauder**, who came from Washington, D.C. in a personal, unofficial capacity to honor this remarkable family. In the front row on the left are **PFC Kelly's** son's mother, **Sandra Berger,** and his son **Doug Berger. (Attachment L p. 330) Forensic Investigator Paul Semones** was also able to join us from New Mexico **via telecommunications.**

We met with PFC Kelly's son's mother **Sandra Berger**; his son **Doug Berger**; and his son's fiancé **Shawn Meagley** in a private dinner setting to honor their loved one. **(Attacment M p. 331)** After a remarkable and memorable three hours, I was able to present the family members in attendance with **plaques** I thought would have been acceptable to the Marines in my Company. **(Attachment N p. 332)** I trust that they are the highest examples of the traditions and esprit de corps of all United States Marines.

It gave me the opportunity to ask for forgiveness from **Sandra Berger; Doug Berger; and Shawn Meagley** for my part in the ordered burial that I consider was the ultimate dishonoring of a United States Marine. Their response was unbelievably gracious. **(Attachment O p. 333)**

The four of us who traveled there together visited the gravesite of **Private First Class Michael J. Kelly** on the morning of November 18th to pay our final respects. In my silent prayer, I asked for his forgiveness. The three of us pictured from left to right are **Col. Jon M. Lauder; me; and John Brown, III. (Attachment P p. 334)**

May **PFC Michael J. Kelly** rest in peace with the knowledge that he has been endeared in the hearts and minds of some of his countrymen and many of his fellow United States Marines for the past 50 years and will be for eternity. **(Attachment Q p. 335)**

In a YouTube release on May 25, 2012, a song entitled "**Semper Fi**" was sung by country/western singer Trace Adkins. He was performing at the United States Marine Corps Base located in Twenty-Nine Palms, California. The lyrics were allegedly written by Kenny Beard, Monty Criswell, and Trace Adkins. I believe they epitomize the all-encompassing meaning of the title of this chapter. They are as follows:

He sat in that long line of barber chairs,
The sergeant asked him,
"Son, would you like to keep your hair?"
He said, "Yes, sir!" as he heard
those clippers buzzing home,
The sergeant said, "Well hold out your hands,
'cuz here it comes."

"Semper Fi!" Do or die,
So 'gung ho' to go and pay the price,
Here's to "Leather-necks," "Devil-Dogs," "Jarheads,"
Parris Island in July, "Semper Fi!"

I sleep in my bed instead of a foxhole,
I never heard my boss tell me to, "Lock and load!"
Ain't no bullet holes in the side of my SUV,
Cuz the kid next door just shipped out for overseas.

"Semper Fi!" Do or die,
So 'gung ho' to go and pay the price,
Here's to "Leather-necks," "Devil-Dogs," "Jarheads,"
Parris Island in July, "Semper Fi!"

For the few that wear the dress blues,
Haircut high and tight,
Who are proud to be the first ones in the fight,
"Semper Fi!"

"Semper Fi!" Do or die,
So 'gung ho' to go and pay the price,
Here's to "Leather-necks," "Devil-Dogs," "Jarheads,"
Parris Island in July, NEVER LEAVE A MAN BEHIND,
You're a Marine, a Marine for life.
"Semper Fi!"

I was able to pay my deepest personal repects to PFC Michael J. Kelly when I visited the Viet Nam Veterans Memorial Moving Wall in Lusk, Wyoming privately and alone on the night of July 13, 2019.

The Betrayal of American Youth

As a young American who believed in serving my Country, I responded positively when I was drafted. It is still very difficult for me to believe I became involved in the burial controversy. It appears to me that I was selectively lied to about my service records as part of the cover-up so that it would be more difficult for me to find the information about the burial of PFC Michael J. Kelly that I have disclosed in this book.

Janet Foster, the staff member in Arkansas Representative Steve Womack's Office, found my Officer Qualification Records in the same Record Center where PFC Kelly's records were located. For several decades, I had been told by the Marine Corps that my records could not be found because they had apparently been destroyed in a fire in Kansas City, Missouri, some years ago. It is very interesting that everything in my OQR seems to be intact...except that there is no information about the burial issue, and only a minor mention about Operation Allen Brook. My OQR also includes documentation that a Certificate for a Silver Star Medal (SSM) was sent to me by the Commandant of the Marine Corps on "21 Jul 69" (July 21, 1969). It was sent to me "Via: Commanding General, 1st MarDiv" (Commanding General, 1st Marine Division). However, my Honorable Discharge from the active Marine Corps occurred on 15 June 1969, and I have never received the Award. I am very interested to know if the reason I have not received the Award is because of retaliation against me for the repercussions generated by the disclosure of the burial of PFC Kelly.

Included in my records is a Letter of Appreciation from the Commanding Officer, U. S. Naval Mobile Construction Battalion ONE, or "Seabees", sent to me as the Commanding Officer of Echo Company while my Company was operating against enemy forces along Highway One over Hai Van Pass. It was forwarded to me via the Commanding General, 1st Marine Division, and the Commanding Officer, 2nd Battalion, 7th Marines. The text of the Letter dated "9 December 1967" said in part, "Your unit's immediate reaction on repeated occasions to aid beleaguered construction crews pinned down by enemy fire undoubtedly reduced many casualties that otherwise would have occurred. Your willingness and ability to furnish immediate support on a moment's notice was outstanding, and your men repeatedly demonstrated that they are truly professionals. Our association with your Battalion, and particularly the men of Echo Company, has provided an inspiring demonstration of the traditional characteristics that have made the Marine Corps great."

An indication of the apparent ongoing cover-up is that, as I mentioned previously, I asked Wyoming United States Senator Mike Enzi to inquire about my missing OQR approximately three years ago. After a period of time, I received what appeared to be hastily copied materials from my supposedly missing OQR that only contained information from my original enlistment date through the approximate time of the 1968 Tet Offensive. While at that time I thought it was odd, it created an absolute determination in me to continue to search for my personnel records of my military service. Apparently, someone in the United States Marine Corps was still not going to allow information to be distributed that would validate or vindicate me in any way concerning the burial of PFC Kelly. Possibly an effort was ongoing to prevent any possible damage to the reputation of the United States Marine Corps... despite the mental trauma I have endured through my life that has almost destroyed me personally and has permanently damaged my quality of life.

The war effort carried out by the United States Government undoubtedly was one of defending South Viet Nam, not defeating North Vietnam and the Viet Cong. The betrayal of American youth was hard to comprehend since young Americans who were willing to serve their country did so, while many refused to or had access to means that allowed them to avoid military service. The book, ***In Retrospect: The***

Tragedy and Lessons of Vietnam, 1st Edition, was written by former United States Secretary of Defense Robert McNamara, who served in both the Kennedy and Johnson Administrations. The book was published in recent years and, as I mentioned previously, it reveals that both Secretary McNamara and President Lyndon Johnson believed as early as one year after the Gulf of Tonkin Incident that their war strategy in Viet Nam would probably fail.

Documents from the Lyndon B. Johnson Library that can be found on the Internet indicate that even the Naval Officers serving on the ships that were supposed to have been attacked ultimately stated in the days following the alleged incidents that the attacks did not occur, but their "eye witness" accounts were ignored. The callous disregard of the irrefutable facts stated by those United States Navy Officers, and the later disclosures in the McNamara book, should be regarded as a criminal offense against all Americans, in my opinion. The revelations of the Arnett interview with Secretary of State Dean Rusk, (See **Introduction** chapter.) if accurate, go far beyond stupidity and ignorance. The actions of at least Secretaries Rusk and McNamara should be regarded as giving aid and comfort to the enemy, which is treason, in my opinion.

An HBO movie entitled "**The Path to War**" that I was able to view in March of 2018 presented another twist on the seeming incompetence of President Johnson and his Administration. I believe the movie, if accurate, is a revelation of the Johnson Administration's concern for the North Vietnamese people in comparison to the apparent disregard they had for the millions of patriots who served the United States in the Viet Nam War. The dismissal of the suffering endured by young Americans who were wounded and/or died there, and those who lost portions of their lives imprisoned there can never be forgotten or forgiven. The billions of dollars of United States taxpayers' money that was uselessly squandered there, and the loss of thousands of United States military aircraft are incomprehensible. The disgraceful manner in which Viet Nam Veterans were treated when we returned home can never be excused. Many Viet Nam Veterans used silence as a form of denial for over 30 years to avoid the discovery of our service to our Country during those tragic years because of the ridicule and discrimination directed at us. Some still remain in the shadows, and mentally stand guard in silence. We now watch in amazement as some of those who either refused to serve our Country, or avoided military service in Viet Nam, attempt to lie and make people believe they served in the Viet Nam War when they did not. On page 404 of the professional masterpiece entitled ***To The Sound of The Guns, 1st Battalion, 27th Marines from Hawaii to Vietnam 1966-1968*** published by BirdQuill, LLC in 2018, the outstanding author Grady T. Birdsong disclosed the following statistical information:

> ***The Department of Defense Vietnam War Service Index officially provided by The War Library originally reported with errors that 2,709,918 U.S. military personnel served in-country. Corrections added 358 not originally listed. Interesting past Census statistics and claims of service in Vietnam found 9,492,958 Americans falsely claiming to have served in-country.***

Those millions of false claims are characterized by many Viet Nam Veterans as attempted "stolen valor." I feel sorry for Americans with those kinds of insecurities and desperate need for recognition. May God help them.

Speaking only for myself, it is easier for me to be left alone and not be used by those who now want to massage their own consciences. All the "Welcome Home Celebrations" will not repair the damage that I have suffered through the years, but I appreciate the efforts of those wonderful people who very sincerely are trying to be of assistance to those of us who served our Country in South Viet Nam. The treatment of the Viet Nam Veterans and the treatment of Veterans in general by the United States Veterans Administration has been a disgrace. I believe those in the federal government who have failed to perform their legal duties to provide services to Veterans should be held criminally responsible. As Americans, we need to stand united in the efforts to ensure American Military Veterans are never treated like we were again. My heart swells with pride and joy every time I see our present Military Veterans honored for their service.

Post-Traumatic Stress Disorder (PTSD)

ON PAGE 151 OF THE BOOK *Tears of a Warrior*, written by Janet J. Seahorn, Ph.D and E. Anthony Seahorn, MBA, the authors included an adjusted quote from another book written by Patience Mason that I believe best summarizes my 50 years of daily experience with PTSD:

> *Veterans lost more than people in war. They lost ideals, belief in their leaders, the innate knowledge that nothing really bad can happen, Some lost hope. Some lost the ability to love or trust or care for anyone. Some become frozen in rage because they feel betrayed by their country and their communities.*

The quote seems as if it were written specifically for me, even though the Seahorn publication by TEAM PURSUITS in Fort Collins, Colorado, has a copyright date of 2008. The following is my abysmal experience with the malady for the past one-half century.

As a result of the efforts of Colorado American Legion Commander Ralph Bozella I described previously, the United States Veterans Administration has certified that I am 50% disabled as a result of PTSD. My post-Viet Nam life has been filled with the agonizing belief that I was somehow different than most other people. I have lived with a compulsion daily to prove that I am adequate as a person, and that I am not a failure. I have tried desperately to conceal my shame of my involvement in the burial issue. During my great childhood and grades 1 thru 12 educational development that I was privileged to experience in a small community in Wyoming, I was always told I was a caring person who had unlimited potential. I experienced more than average popularity among my peers and had expectations that "the sky was the limit" as far as achievement in my personal life.

I could not accept, until I was in my 60s, that somehow, I was different than I had been predicted to be by my parents and adult mentors of my youth. When I returned from South Viet Nam to the sudden isolation of rural Wyoming for my 30 days of leave prior to going to Hawaii, I felt paranoid and I was desperate to be able to communicate with someone who I had just spent 13 months with. I was engulfed by indescribable mental trauma concerning the burial issue. It seemed like people were almost afraid of me for some reason, and they appeared to not want to engage me in any personal manner. Maybe there was a change in my personality that was evident to the people I had known all of my life.

I constantly attempted to conceal my true emotions. I desperately did not want anyone to regard me with the apprehensions that our family did when they were around my WWI Veteran Great-Uncle Guy Hanna. The sound of the "chopping" noise produced by the two blade Huey helicopters that are still being flown at Warren Air Force Base in Cheyenne, Wyoming, instantly sends "chills down my spine," and I immediately feel a sense of urgency and heightened awareness. I often avoid 4th of July Celebrations because the popping noise and explosions of the fireworks sometimes make me have "flashbacks."

In 2017, I experienced a flashback during the annual 4th of July Celebration that is always held at the small town of La Grange, Wyoming. I live about one and one-half miles from town, "as the bird flies," and there is a hill between my small ranch and the celebration site. As darkness set in that night, I was using my tractor to place fill dirt around some of the livestock watering tanks in my pasture. When the first lights of the fireworks began to flash in the air and the noises from the explosions began, my livestock suddenly bolted in fear and ran to the border fence approximately three-quarters of a mile away. Instantly, I was consumed with fear and temporarily my mind was filled with memories of death and destruction. It was a reminder to me that

I still need to remain vigilant of my surroundings daily to help control my reactions. Attending funerals has always been an agony beyond description for me, but I constantly try to not let it be noticed.

I was part of a horrible marriage in which I was reminded very frequently that I was unacceptable and that every negative issue in my marriage was because I was "so messed up from being in Viet Nam." Maybe that is correct. The inadequacy I felt over the burial of the Marine drove me constantly to attempt to prove my worthiness in every situation, but it seemed like I never could. The sense of failure haunted me nearly every hour of every day, and I was driven to always be doing something, so I did not have time to think about how I had failed every American who had ever served in the United States Military.

The shame I still feel is so intense that I constantly struggle with thoughts of how meaningless my life has been because of the burial. For many years, I did not sleep over four hours a day. I self-medicated with alcohol for a number of years, much of which was done in the privacy of my home when I spent so many years virtually alone. When coaching high school students, I had constant flashbacks that instantly would make me relate a mistake by a young athlete in a basketball game as being a matter of life or death, which was my real life experience with young men just a couple of years earlier.

To me, danger lurks around every corner and behind every closed door. It is often so intense that I constantly have to be on guard to control my actions, reactions, and emotions. I never feel safe in any situation, and I always try to have defensive safety plans that include some kind of safety weapon at my disposal. Sudden loud noises cause a visceral reaction in me that I struggle to control, and somehow, I have a sensory perception that in the microsecond before a loud noise occurs, I hear the noise coming. I always feel compelled to sit with my back to the wall as much as possible, and enormous trust issues plague me in personal associations I have with others and the relationships I engage in. When someone approaches me from behind, and I am not aware they are coming, I often react and recoil physically when I sense they are there, both night and day. My feelings of failure were accentuated by my financial loss of my farm and ranch operation and my trucking business as a result of my bankruptcy brought on by a divorce property settlement that I could not pay in the 1980s.

The attention PTSD has received in recent years has been very important in the lives of many Veterans, but it has also caused many who suffer from the disorder to stay in the shadows because of the possible denial of Second Amendment Rights, the inability to obtain certain employment, and the possibility of being ostracized because of suffering from mental health issues. The actual events that create PTSD vary, as do the symptoms, but the agony for many who are afflicted is indescribable. For the average person, the process used to determine if someone is suffering from PTSD, and to what extent, is not clear. How is the percent of PTSD disability determined? What is the difference between being disabled to a certain percent, or being considered totally disabled from the disorder?

While I do not know the answers to those questions, I can only say that after almost 50 years, my affliction is real. However, I have discovered several of my "triggers." I have taught myself some ways to attempt to deal with it when I feel myself acting or reacting, and I have learned through the many years to live with it, and somehow manage my life. However, the loss of loved ones and failed relationships with friends who displayed boundless caring, kindness, compassion and love toward me through the many years prompted my decisions to try to avoid potentially positive experiences because of my inability and failure to believe in myself.

In my readings about and study of PTSD through the years, and with the advice of professionals in the field of mental health who were personal friends of mine, I have been able to realize that there can be significant support for those suffering from a trauma when they confide in family members. I attempted to do that with one of my daughters at one point a few years ago, and as soon as I breached the subject that I was considered to be partially disabled by the Veteran's Administration due to PTSD, she very dismissively and simply said, "Does that mean you will get a check?" Needless to say, I did not mention it to anyone again. By contrast, I have two kind and considerate stepdaughters that I helped raise during a second marriage that, unfortunately, also ended in failure. They have always been very kind by thanking me for my military service and recognizing my Veteran status, and for that I will be forever thankful.

As strange as it may seem, one thing that has helped me deal with my PTSD has been my years of public service to others. The responsibilities of the positions I have held, and the scrutiny of the public image that accompanies those positions, have given me courage to continue at times when I have felt there was nothing left to live for. I thank God daily for having had those opportunities.

In summary, the personal and sociological effects of PTSD on me, the issues behind the on-set of it, and the effects of it on my life, can be best summarized by an often quoted simplicity: "It was what it was, and it is what it is." Tomorrow is yet another day to continue on.

In Retrospect

NOTHING IN THIS BOOK should be interpreted to be an attack on the United States Marine Corps as an institution. I was not a Marine, I am a Marine. I will always be proud of that fact. However, as in any organization, the despicable actions of the "few" can always reflect negatively on the outstanding, positive efforts of the "many." The burial issue I have related to you is one of those tragic situations.

At the conclusion of this writing in 2019, United States Marine Corps Colonel Jon M. Lauder is the Director of the Liaison Office between the Commandant of the Marine Corps and the United States Senate. Colonel Lauder is the epitome of the outstanding quality of Marine leadership in the present-day Headquarters, United States Marine Corps. With leaders like him, I believe the Marine Corps is steadfast and in an excellent position for continued greatness.

I appreciate the efforts of our younger generations of Americans to bring honorable recognition to Viet Nam Veterans. It is interesting to me that many of those who would not serve their country in the military during the Viet Nam War still lurk silently in the shadows, and many now even make false claims of their service to our Country in South Viet Nam to try to somehow be socially acceptable. Those Americans who refused to serve and manipulated the Selective Service System should be apologizing publicly for their actions, not leaving it to the younger generations to try to atone for the actions of "the privileged" who have disgraced Viet Nam Veterans for decades.

The most important thing needed by Viet Nam Veterans is the support of their families and loved ones. While the trauma that has surrounded me through the years could have made me forget, I do not remember being thanked by my parents, any of my siblings, or my children for my service in South Viet Nam. I do not recall any of them saying that they were proud of my military service to our Country. Again, while I may have forgotten, I also do not remember any member of my immediate family wishing me a Happy Veteran's Day with one exception. One of my sisters has started recognizing me on Veteran's Day for about the last ten years.

The burial issue has obviously been a source of agony for me through the years, and I felt compelled to make an effort to contribute to the national dialogue when the United States Marine Barracks were destroyed by a truck bomb in Beirut, Lebanon, on October 23, 1983, during the Reagan Presidency, 220 United States Marines, 58 French military personnel, 21 other military personnel, and six civilians were killed in the atrocity. My Officer training was explicit that, if at all possible, you should never mass troops in a war zone. The obvious reason is that, if you allow that to happen as a leader, you have created a target of opportunity for possible enemy activity. Some prognosticators have surmised the Marines were housed in that facility in an apparent attempt to make the United States military presence in Lebanon appear to be more casual and less militant. However, there were approximately 1,800 U. S. Marines in the country at the time. The Marines killed there were the victims of horrible leadership decisions within the Marine Corps, in my opinion.

I wrote to President Reagan and suggested that everyone in the Marine Corps who had participated in the command decision to house those Marines in that building should be relieved of their duties, from the Commandant of the Marine Corps on down the chain of command. I expressed in detail my concern that the tragedy would be covered up, just as the Marine burial in South Viet Nam had been. Approximately ten days after I sent the letter, I received a call one afternoon, and the person calling said the President had received my letter and that he would call me to discuss the contents of the letter that evening at 7:00 p.m. However, the phone never rang. In my thoughts, and possible imagination, it appeared to me that, once again, the sordid burial issue was covered up.

Maybe the "curtain of silence" surrounding the burial issue over the decades of time has existed because of a lack of respect by those in the higher echelons of Command in the Marine Corps for those actually involved in the burial. Possibly it has been considered "no big deal" by individuals who have been made aware of it since that time. Hopefully, my mental health, and that of many of my men who have endured the trauma created by this indescribable tragedy for over one-half of a century, will be improved. The many facts that have been discovered have finally answered many of the numerous questions that have haunted many of my men in Echo Company 2/7 and me for one-half a century.

Spiritual leaders who helped shape my religious beliefs in my adolescent youth have shunned me through the years, even when I desperately sought their assistance. The few who ever spoke to me simply quoted the biblical passage that they thought people who had the same spiritual beliefs I did should adhere to; "Render unto Caesar what is Caesar's." They also advised me that the reason I was struggling emotionally was because "I had sin in my life." I have always chosen to believe that they did not understand the scope of what I was dealing with, and I still hope today that somehow my belief is true.

Epilogue

THIS IS MY PERSONAL ACCOUNT of my experience in the United States Marine Corps primarily during the Viet Nam War from July 1967 until August 1968. While the passage of time and the "fog of war" may have made my reflections and memories not as accurate as I might wish, the passage of time has allowed me to view the overall circumstances in a more perceptive way that hopefully makes this writing more understandable and reasonable to those who read it. I have taken the time to engage most of the people in this book, whose names are disclosed, to substantiate my recollection of the events related herein, and as verified by the research of the military records that have been found. Dr. Donald Catherall published his memoir, *Shaped by the Shadow of War*, in Lexington, Kentucky on August 3rd, 2017, and I have quoted it extensively to substantiate my recollections.

I have attempted to have verification of the historical facts and dates contained herein authenticated. I am grateful to Doug Cubbison, Historian for the Wyoming Veterans Museum, for his encouragement. In addition, I could not have completed this book without the military judgment, professional sophistication, and lasting encouragement of United States Army Lieutenant General Edward Wright, (Retired), and United States Army Major General Charles Wing, (Retired). General Wright culminated his military career by dedicating eight years of his public life to serve as the Adjutant General of both the Army and Air Force National Guard forces in Wyoming, as General Wing did prior to that. I will be eternally grateful for their guidance and assistance, and the personal confidence they have entrusted in me in the writing of this book.

In addition, the detailed critiques, accurate reflections, and thoughtful suggestions of fellow Viet Nam Veteran and Marine Officer Colonel Andrew Blenkle, United States Marine Corps (Retired) has been irreplaceable. His work in support of my writing of this book was that of a true friend and dedicated Marine Corps Officer. Tragically, Colonel Blenkle lost his struggle to a terminal illness in 2018.

John Brown, III, former President of John Brown University and the Executive Director of the Wingate Foundation, had been one of the cornerstones of my spiritual faith throughout my life. He formally contributed support and encouragement to me in a way that allowed me to carry on, when it appeared to me that I would never be able to finish this writing.

Marine Colonel Jon M. Lauder, who was serving as the Commandant's Liaison to the United States Senate in 2018-2019, played an indispensable role in my efforts to complete these writings. He distinguished himself in combat leadership in both Iraq and Afghanistan, and he is the pinnacle of American Military Leadership moving forward to lead the United States military in the perilous future of our Country.

My recognitions must include further comments concerning Lt. Jim Pippen. His presence as a Platoon Commander on my Echo Company Staff in South Viet Nam was solid and steadfast. He is a true southern gentleman in the finest meaning of the word. He never "buckled" under pressure, and he had the self-confidence and courage to never blame others for events that transpired around us. He is one of those individuals who I could always rely on and, in my opinion, meets the qualifications of the phrase we have in Wyoming, "The World Needs More Cowboys." Jim is a "Cowboy" in my opinion, and I could not have finished this book without his faithfulness as a fellow Marine and friend, and for his spiritual support.

My acknowledgements can never be complete because of the numerous individuals that have contributed immensely to the writing of this book. Two excellent Marines who were outstanding Americans serving in South Viet Nam with me in Echo Company were Corporal Joe

Freda and Corporal Joe "Hawk" Wnukowski. Their ideals and personal integrity exceed the very highest traditions of the United States Marine Corps. The historical preservation of those minutes, hours and days surrounding the discovery of the remains of PFC Kelly, including capturing the emotions of all Marines with the picture that adorns the cover of this book, can never be forgotten. I have distinct pride that I was able to have had them under my command. A simple "thank you" is not enough for these fine men. **In addition, I salute them and honor them.**

This book would not be complete without me paying tribute to some of the people associated with John Brown University who helped reinforce and expand the foundations of my spiritual values and afforded me friendships that will transcend my life:

Dr. John Brown, Jr. was the President of JBU when I was recruited to play basketball there after I left the University of Wyoming. Dr. Brown was one of the finest people I ever met. His positive leadership, reinforcement of the value of the lives of everyone he came into contact with, and his vision for the role JBU could play in the success of every student were matchless;

Reverend Jerry Hopkins was the Campus Minister at JBU, and he was a spiritual leader beyond compare;

Dr. Mabel Oiesen was a Fine Arts Professor and Director of the Cathedral Choir who, even at an advanced age, had the ability to touch the heart and mind of every student she came in contact with;

The many JBU faculty and staff who provided leadership, love, and caring;

Murray Turner is the most talented singer and vocal musician I have ever known, and he was one of the leaders of the Harmonaires singing group I was a member of while a student at JBU. Murray exemplified the genuine person who lived what he believed spiritually, everyday;

Roger Cross later became a minister. His everyday life, his continual commitment to his spiritual convictions, and his leadership were recognized internationally as personified by his being elected President of Youth for Christ International for many years;

Bob Jones grew up in a small farming community in Missouri just as I had in Wyoming. We met as teammates on the JBU basketball team, and our friendship endured throughout our adult lives, which included a lasting long-distance friendship between our mothers;

The entire JBU family including the student body from 1962-1964;

One student in particular taught me what a true, pure, personal relationship is when it is based upon mutual personal and spiritual values, and absolute, all-encompassing respect and love for each other. My experience of that individual relationship has always inspired me to continue to try to live another day, even during my darkest hours;

Archie Johnson is a JBU Graduate who was my childhood and life-long friend in Wyoming, and who is still a neighbor near La Grange, Wyoming. He had an asthmatic condition that prevented him from serving in the military, but my dad often spoke of how much "Arch" helped him while I was in the military service, and how much it meant to him. "Arch" is one of those unique individuals who I have never heard anyone say a negative thing about;

Paul Semones graduated from JBU in 1999, with a B. S. degree in Mechanical Engineering, after which he founded Semones Forensic

Engineering. He entered my life in 2018 at a critical time during the writing of this book. Without his professional talent, this book would not have been completed in the specificity and with the documented proof that it contains;

John Brown, III, grandson of the founder of JBU, has been an immensely dependable person I have been able to turn to through the years, and he has instilled confidence in me to carry on, even though I am sure he has not been aware of his profound influence on my life.

I only hope that all of those at John Brown University I have known through the years will somehow realize how grateful I am to, and thankful for, them. Without the guidance and immense personal influence on me in those formative years at JBU, the foundations of my very existence would have crumbled later, and I would not have survived to this point in my life. The head, heart, and hand training and learning philosophy of JBU is the foundational cornerstone in the lives of those who graduate from that hallowed institution.

I have experienced great apprehension about this written revelation of my life. To reveal my personal life that I have kept "camouflaged" for so many years has been terrifying for me. However, when I decided to make the commitment to put a very personal part of my life in writing for the many reasons I have tried to explain, I sought to reveal those things that I hope will be the most pertinent, and possibly helpful, to anyone who may read this book.

Personal Biographical Information

Enoch Baumgardner enlisted in the United States Army in 1966. He was meritoriously awarded the opportunity to attend Officer Candidate School from which he graduated in 1967 as a Second Lieutenant. After he completed his tour of duty in Viet Nam in 1969, he returned to civilian life and graduated from Eastern Wyoming College in Torrington, Wyoming and later attended Chadron State College in Chadron, Nebraska where he pursued a Business Degree. He then returned to Goshen County, Wyoming where he has maintained a renowned ranching business since that time.

Max Beebe graduated from John Brown University in 1963. He joined the United States Army and earned his "wings" as a helicopter pilot. He was deployed in South Viet Nam with the 188th Helicopter Company, known as the "Black Widows," in 1967. After returning from his Viet Nam Tour, he was stationed in Germany. In 1971, he returned to South Viet Nam as a member of the 120th Assault Helicopter Company, known as "The Deans." He was chosen to fly military leaders, including General Abrams, throughout the Viet Nam Theatre of Operations. After attaining the rank of Major, Max left the military for a short period of time but returned to military duty with the Wyoming National Guard as a Chief Warrant Officer so he could continue to pursue his passion of flying. His missions included the distribution of medications and drug education programs in public schools, and a broad spectrum of other services for the State of Wyoming. He retired from his service in the Wyoming National Guard and the United States Army in 2003 after 35 years of service to our Country and the State of Wyoming.

Mike Bryant enrolled in John Brown University in 1961. The Oklahoma native was drafted into the United States Army in 1966, but he was allowed to transfer to the United States Marine Corps where he was selected for enrollment in the MARCAD program for enlisted Marines as a Candidate. When he successfully completed the MARCAD training, he became a helicopter pilot even though he had not completed his college degree. He was stationed in Viet Nam from September 1967 until October 1968. After his completion of his obligation to the Marine Corps, he returned to JBU and completed his degree in Education graduating in 1971. He later retired from Chevron Oil Company after 25 years of employment in oil pipeline operations.

Donald Catherall was my Radio Operator from the day I became the CO of Echo Company until the day he rotated home... just three days before Echo Company was deployed on Operation Worth. Don went on to obtain a Ph.D. in psychology and had a distinguished career in the field of posttraumatic stress. He published several books on the impact of trauma on families, served on a VA national advisory committee on combat veteran's readjustment, and has been teaching the treatment of trauma disorders at Northwestern University since 1985. Recently, he self-published a memoir of his experience in Echo company in Vietnam and his father's experience in 1st Marines at Peleliu in WWII. Don's daughter is a political consultant and his son is a fly-fishing guide in Idaho. At this writing, Don and his wife are both still practicing in Chicago.

Joseph W. Freda enlisted in the United States Marine Corps in 1966. Upon completion of Boot Camp, he was deployed to the U. S. Marine Corps Base Guantanamo Bay, Cuba until 1967. Later in 1967, he was deployed to South Viet Nam. When he arrived, he was assigned to Echo Company, 2nd Battalion, 7th Marine Regiment, 1st Marine Division. Upon the completion of his 13-month tour of duty, he was stationed at Cherry Point Marine Corps Air Station, North Carolina where he was a rifle instructor. Joe completed his enlistment in the Marine Corps in 1969 as a Sergeant. He returned to Daytona Beach, Florida and was a fuel delivery driver for his father's fuel business. He completed training as a home and commercial boiler installer and became a master pipe fitter. He started an installation company and then assumed management of his father's business. After 18 years of business management, he sold the business and became the service manager for a federal heating and engineering firm. He entered retirement from that position 19 years later. Joe carries the sorrow of the loss of his brother in 1966 who was serving his Country in Viet Nam as an Army Crew Chief in a Huey helicopter when it was shot down by enemy forces.

John C. Johann was transferred to the Fleet Marine Corps Reserve on 1 June 1988, and subsequently was retired from the United States Marine Corps as a Master Sergeant on 1 June 1991 after serving 30 years in the regular and reserve components of the Marine Corps. Upon retirement, he became employed by the United States Marshal Service as a Special Deputy United States Marshal and continued his employment there until he retired again in 1998. He is presently enjoying retirement in the foothills of the beautiful Bitterroot Mountains of Montana.

Jon M. Lauder is, at the time of the writing of this book, a Colonel in the United States Marine Corps and serves as the Liaison between the Commandant of the United States Marine Corps and the United States Senate. Colonel Lauder has a very distinguished record of service to the United States of America, including deployments with the 2nd Tank Battalion in support of Operation IRAQI FREEDOM and the Command of the 2nd Tank Battalion when it was deployed in support of operations in Helmand Province, Afghanistan. He has also graduated with distinction from the Top Level School at the National War College. The Defense Meritorious Service Medal and Meritorious Service Medal with three gold stars in lieu of third and fourth awards are among the numerous awards and honors he has earned during his service to our Country.

John Reily Love was born October 15, 1929, in Memphis, Tennessee. He grew up in Arizona and graduated from the United States Naval Academy in 1951. Second Lieutenant Love was assigned to the United States Marine Officers Basic School, Quantico, VA prior to his deployment to Korea. He participated in the 1952 Korean War Spring and Summer Campaigns. Upon his return, he was stationed at the Marine Barracks, Washington, D.C. where he was promoted to First Lieutenant in 1952, and then to Captain in 1954. He entered the Amphibious Warfare School as a student in 1961, and was promoted to Major in 1962. In 1965, he was selected to serve as Aide-de-Camp to the Assistant Commandant of the Marine Corps in Washington, D.C., and was promoted to Lieutenant Colonel in 1966. In 1967, Lt. Col. Love was ordered to Vietnam where he commanded the 2nd Battalion, 7th Marines from August, 1967 until February, 1968 after which he was assigned to be the Executive Officer of the 7th Marine Regiment of the First Marine Division. In 1971, he was transferred to Camp Pendleton, CA and assigned to be the Commanding Officer, Headquarters Battalion. He was promoted to the rank of Colonel in 1972. In 1977, he was ordered to attend the United States Army War College, Carlisle, PA as the Marine Corps Representative and faculty member where he remained until his retirement from the Marine Corps in 1981. Upon his retirement, Colonel Love became the Operations Manager for the Vinnell Corporation in Saudi Arabia until 1985. Colonel Love then devoted his time to serving as an Adjunct Faculty Professor for four New England colleges. In 1995, he moved to Carlsbad, CA where he was deceased in 2010.

Ronald J. Maines graduated from John Brown University in 1966. After his initial obligation as an officer and helicopter pilot in the United States Marine Corps, he reentered civilian life. As a pilot, he led Mission Aviation Fellowship's largest program in Indonesia and later led relief efforts for World Vision International in Asia and Latin America. His leadership was exemplary in business development and humanitarian efforts which included an orphanage program in Romania and World Vision leadership assignments in the Philippines,

Vietnam, Afghanistan, and the United States. Having earned a graduate degree in Systems Management from the University of Southern California, he has also completed studies at the Peter F. Drucker Graduate School of Management. Ron is semiretired in Rogers, Arkansas but still serves as the Principal of Crossroad Leaders: Helping Leaders Invent The Future.

Jimmie R. Pippen graduated from Northwestern State University in 1967, and joined the United States Marine Corps. After his release from active duty in 1970, he was employed in business management until he founded Custom Heat Treating, Inc. in Bossier City, Louisiana in 1975. After selling that business, he owned food service and photo injection businesses. He finished his business career with Southern Heritage Life Insurance Company in Jackson, Mississippi until he retired in 2008.

Gabriel R. Quinones, the son of a World War II Veteran, left the United States Marine Corps at the end of his enlistment. He eventually became employed by the Wonder Bread Corporation and retired as a member of the Teamsters Union. He and his wife Lois raised four children, two of whom served our Country in the United States Air Force and the United States Navy. Gabe has suffered many years from the effects of our exposure to Agent Orange in South Viet Nam. Numerous cancers in various organs have resulted in him being classified by the United States Veterans Administration as being 100% disabled.

L. C. Stair, III graduated from Yale University in 1966, and joined the United States Marine Corps. After his release from active duty at the end of his Reserve Officer obligation to the Marine Corps, he was admitted to the Tennessee College of Law and received his law degree in 1972. He spent his legal career with the law firm of Bernstein, Stair, & McAdams in Knoxville, Tennessee, and retired from the practice of law effective January 1, 2017.

Joseph T. Wnukowski was born February 12th, 1948 in Philadelphia, PA on Abraham Lincoln's Birthday. His mother had escaped from Poland in 1939 amidst the ravages of WWII, and finally found safety as a refugee in Tanzania. Because of his Polish heritage, his father, an American citizen, was sent to Africa by the U. S. Government to assist Polish refugees. He met and married Joe's future mother there. She narrowly avoided the travel restrictions for pregnant women trying to reach the United States that was enforced because of the threats posed by German U-boats. Joe enlisted in the Marine Corps in 1966, served in Viet Nam as a member of Echo Company, Second Battalion, Seventh Marines during 1967-1968, and left the Corps as a Sergeant in 1969. He was employed as a High Voltage Lineman by the Philadelphia Electric Company and retired after 43 years of service.

ATTACHMENTS

19

Our Servicemen in Vietnam
(Defense
Presidential Handling)

A2/CS

RECEIVED
1968 APR 8 AM 14 27
OFF SECY OF DEFENSE

Cpl. Willie Williams
2/9 Hd. Qt. 1st Marine
% FPO. San Francisco
96602.

Mr President
Whit house
Washington D.C.

Mr. President;
I'm a Corporal in the united States Marine Corp. We are out on an opperation North west of ten kee. the name opperation worth.
Today Mr President we found a Marines body wraped up in a puncho laying in a bomb crater. We brought it to our

C.P. (Command Post) and our Capt. Chamberlan inform his commanding officer and they told him to bury him so what could he do sir, But bury him. Sir Marines don't supposed to be buried in this Country plus I think his unit ran off and left him and are now trying to cover it up. Please look into it.

unforgetable yours
Cpl. William
Mr. President

PS

Mr. President
Men was sent in for this missing Marine, they brought back something and they got the bronze star.

Look into it Sir.

Please Look into it

CASUALTY ASSISTANCE CALL REPORT (1770)
NAVMC HQ 362 (REV. 11-65)
SUPERSEDES NAVMC 10149-PD WHICH IS OBSOLETE AND WILL NOT BE USED.

DATE: 2 Apr 1968

FROM: (Zone and station of officer making call)
1stLt G. A. RAVAN 097690 USMC
Inspector-Instructor Staff, OrdMaintCo(-), MaintBn, ForTrs, FMF, USMCR, MCRTC
1151 W. Buckeye Road, Lima, Ohio 45804

TO: Commandant of the Marine Corps (Code DNA), Headquarters U. S. Marine Corps, Washington, D. C. 20380

REFERENCES: (a) CMC letter dated 29 Feb 1968 to XXXXXXXXXXXX, I-I
MaintBn, 4thFSR, ForTrs, FMF, USMCR, MCRTC, Lima, Ohio
re Casualty Assistance Call, case of Pfc Michael John KELLY 2342397 USMC
                                                                    deceased.

(b) MARCORPERSMAN, Chapter 12, Part F

1. In compliance with references (a) and (b), the following report in duplicate is submitted herewith:

ASSISTANCE RENDERED

A. DEATH CERTIFICATES
Explained the meaning of the certificates and advised NOK to retain in a safe place.

B. BURIAL ALLOWANCES, REIMBURSEMENT FOR EXPENSES; GOVERNMENT HEADSTONES; MEMORIAL FLAG
Completed request for payment of burial allowance. Headstone (flat bronze marker applied for; memorial flag presented at funeral.

C. ARREARS OF PAY
Applied for and explained to NOK.

D. DEATH GRATUITY
Applied for and explained.

E. PERSONAL EFFECTS
Explained method of shipment and the normal delay time to be expected.

F. DEPENDENTS ASSISTANCE AND ALLOTMENTS (Explanation for discontinuance of)
Explained discontinuance of allotments.

G. TRANSPORTATION OF DEPENDENTS
Not applicable.

H. TRANSPORTATION OF HOUSEHOLD GOODS
Not applicable.

I. DECORATIONS AND AWARDS
Explained that they would be forthcoming and that the undersigned officer would make arrangements for presentation.

J. U.S. GOVERNMENT LIFE INSURANCE
SGLI in effect and assisted NOK in completing application.

FILE ELW

Attachment B

1. NATIONAL SERVICE LIFE INSURANCE
   Not in effect.

2. DEPENDENCY & INDEMNITY COMPENSATION
   Explained and assisted in completing forms.

3. SOCIAL SECURITY
   Explained and assisted in completing forms.

4. COPIES OF DEATH CERTIFICATE (CIVILIAN)
   Explained the use of death certificate copy for claim of insurance.

5. BURIAL FLAG - BURIAL ALLOWANCE
   Advised.

6. MILITARY SERVICE CONNECTED PENDING VA CLAIM
   Not applicable.

7. HOSPITAL & MEDICAL CARE
   Not applicable.

8. EXCHANGE & COMMISSARY PRIVILEGES
   Not applicable.

9. EMPLOYMENT
   None requested. (Not applicable.)

10. AID (NON-FINANCIAL ASSISTANCE)
    None requested. (Not applicable.)

11. EDUCATION
    Not applicable.

12. LAST RITES BURIAL
    Advised as to policies of the Society and nearest address.

13. TRANSPORTATION AND/OR HOUSING
    None requested.

14. OTHER PROBLEMS OR COMMENTS
    None.

Attachment B-2

1. CHANGE OF ADDRESS OF NEXT OF KIN
NOK advised to keep CMC informed of any change of address in the future.

2. Do you recommend that the dependent's case be referred to the Navy Relief Society for assistance pending payment of such claims as may be due? ☐ YES ☒ NO

3. FURTHER COMMENTS, OBSERVATIONS AND RECOMMENDATIONS *(See back of this sheet if additional space is required)*

Informed NOK on 19 February 1968, that their son Pfc Michael J. KELLY was KIA on 15 February 1968 in vicinity of Quang Nam Province Republic of Vietnam result gunshot wounds to the body from hostile rifle fire and fragmentation wounds to the body from a hostile grenade while on patrol. Family accepted the loss of their son as the will of god and were proud that he served with the United States Marines doing a job in Vietnam that must be done. All possible assistance was rendered, and arrangements were made pending arrival of the remains.

Due to identification difficulties, the remains were not received until 27 March 1968, this long delay resulted in extreem emotional strain on the family. Their concern was accelerated when a letter was received from Pfc KELLY's Plt Leader indicating the remains were recovered intact, this situation prompted a member of the family to seek congressional assistance in expediating the return of the remains. Despite these circumstances, the family was at all times understanding and grateful for all assistance rendered.

Pfc KELLY was buried with full military honors on 28 March 1968.

No further complications were encountered.

G. A. PAVAN
(Signature of Officer)

DATE ORDERS RECEIVED: 29 Feb 1968
DATE CARD RECEIVED: Not received
DATE CALL COMPLETED: 1 Apr 1968
MILEAGE ONE WAY: 44 miles one way

Attachment B-3

Letter from 2Lt Mark C. Whittier, platoon leader, C/1/7, regarding death of PFC Michael John Kelly

Received from Doug Berger (Kelly's son) on 5/12/2018 as four image files

Transcribed by PTS into Word on 5/12/2018

Begin:

(Page 1)

2 March

Dear Mr. & Mrs. Kelly,

Please excuse this paper, but I just received your kind letter and, as it is late in the evening I was unable to get paper. This is insignificant, however, but I am glad you decided to write me back. As a platoon commander I would so much enjoy and, at times, would need to write parents about their sons over here. But time just doesn't permit this and the only time I can find time is when something like this has happened. So I will be as lengthy as time permits in order to fulfill your request and to be able to speak to the wonderful people who made my Marines.

Mike's squad was the point or leading squad of a battalion size column that was moving through the mountains in search of the Viet Cong. We had just passed another Marine unit that had been hit and lost 4 men, so we knew that the VC were around. I chose 2nd squad because they were the best of my 3 squads and Mike was the point or lead man. He was really sneaky on

(Page 2) his feet. He had one of the best senses of direction I've ever seen and could almost smell the VC when they were near. He saved his squad on 2 occasions from ambush. That would have been deadly. He can be credited for many sightings and kills [hills?] while on patrol as a result of his alertness. I felt that he and his squad would be best suited and most reliable to lead a large operation like this one. In a way I feel guilty for my choice, but then again it was a necessary one and I could only choose the best with the responsibility I had. We may have some consolation in knowing that no more men were lost than we did lose.

As we were moving down this path I heard shots from an M-16 rifle that I learned later was Mike's. But later we found that his firing was only in vain for the VC were heavily dug in. They opened up with heavy machine gun fire from heavily concealed and fortified bunkers. He died instantly and was 20 meters in front of the second man. (He always made the 2nd man stay that far behind in case he hit a mine

(Page 3) that might hurt 2 men). One of the rounds hit a white phosphorus grenade used for target marking and smoke screens and burnt him badly. As I said before he most likely knew nothing of what was happening.

We lost two more men but got his body out shortly afterwards. We removed it to a spot where we could get him out the next morning by helicopter. The next morning we tried to get him out and we

Attachment C

were attacked by the VC who pushed our platoon down off the hill. We sent the 8 man squad up to get him that night and his body was safe within our lines within 18 hours after we had left it.

We had a memorial services [sic] for Michael and the others about 3 days afterward and the sermon that the chaplain gave was very inspiring and encouraging. It all boiled down to the so true fact that these men and Mike all died so that their children or families would not have to endure the pains and suffering that we in Vietnam are having to endure. I am proud to say, and I'm sure you will concur, that Michael died willingly and with the love of his family in his heart.

All his squad joins me in again conveying

(Page 4) our sincerest wishes that the days to come will be easier for you. You have given a mother's most prized possession – her son. But you can certainly be proud of Michael. He was worth his weight in gold as a Marine too.

Please ask me again if there is anything else I can do to help you. Thank you again for the letter. And best regards from 2nd Platoon.

May God bless your whole family

Mark C. Whittier

[PTS note – 2Lt Mark Crosby Whittier, C/1/7, was killed in action 4 days after this letter on March 6, 1968.]

7 March

Dear Mr. & Mrs. Kelly,

Please excuse this paper, but I just received your kind letter and, as it is late in the evening I was unable to get paper. This is unimportant, however, but I am glad you decided to write me about it. As a platoon commander I would so much enjoy and, at times, would need to write parents about their sons overseas. But time just doesn't permit this and the only time I can find time is when something like this has happened. So I will be as lengthy on this points in order to fulfil your request, and to be able to speak to the wonderful people who were my marines.

Mike's squad was the point or leading squad of a battalion-sized column that was moving through the mountains in search of the Viet Cong. We had just passed another mine unit that had been hit and lost 4 marines, so I knew that they were around. I chose his squad because they were the best of my 3 squads and this was the point or lead man. They were really moving on his fire. He had one of the best senses of direction I've ever seen and could almost smell the VC when they were near. He saved his squad on excursions from ambushes that would have been deadly. He soon he credited for many sightings and kills while on patrol as a result of his alertness. I felt that he and his squad would be best suited and most valuable to lead a long operation like this one. Also in a way I felt guilty for my choice, but then again it was a necessary one and it would only show the best with the responsibility I had. This may have some consolation in knowing that no more men were lost than we did lose.

As we were moving down the path I heard shots from an M-16 rifle that I learned later was Nahl's. But later we found that his firing was only in vain for the VC were heavily dug in. They opened up with heavy machine gun fire from heavily armored and fortified bunkers. The [men] instantly ran for cover — first of the second [man]. We always wait [for] and men say that few [soldiers] in war he hit a sure that night (about 10 men). One of the wounded hit a white phosphorus grenade used for target marking, and smoke screens and burnt him badly. As it went before he most likely knew nothing of what was happening.

We lost two more men but got the body out shortly afterwards. We carried it to a spot where we could get him out the next morning by helicopter. The next morning we tried to get him out and we were attacked by shells who pushed our platoon down off the hill. We sent the 8 man squad up to get him that night and his body arrived within our lines within 1 hour after we had left it.

We had a memorial service for Michael and the others about 3 days afterward and the sermon that the chaplain gave was very inspiring and encouraging. It all boiled down to the sober fact that these men and others all died so that their children or families would not have to endure the pains and suffering that we in Vietnam are having to endure. I am proud to say, and I'm sure you will concur, that Michael died valiantly and was the best of his family to his death.

All his squad gave me an again assurance

our sincerest wishes that the days to come will be easier for you. You have given a mother's most prized possession – her son. That you can certainly be proud of Michael. He was worth his weight in gold as a marine too.

Please let me know if there is anything else I can do to help you. Thank you again for the letter. And best regards from Ed Hutson.

May God bless your whole family

Mrs. L. Whitten

10 April 1968

Dear Corporal Williams:

President Johnson has asked that I reply to your recent letter.

In view of the situation described by your letter, I certainly understand your concern. In this regard, I have learned that the Commandant of the Marine Corps had already been apprised of allegations similar to those presented by you and that action has been initiated to conduct an appropriate inquiry. Accordingly, I am forwarding your letter to Headquarters, U. S. Marine Corps, for information and whatever further action is considered necessary. I trust that I have acted in the best interests of all concerned.

You may be assured that the President and all of us in the Department of the Navy share your understandable concern in this matter.

Sincerely,

Charles A. Bowsher
Assistant Secretary of the Navy

REPLY COORDINATED WITH OSD — COLONEL MURPHY by telecon
Case of: Cpl Willie Williams, USMC, 2287883

Cpl Willie Williams, USMC
"E" Company 2nd Battalion 7th Marines
1st Marine Division
FPO San Francisco, California 96602

Prep by: CAPT OSTER, USMCR, WHCS, X-50191
Typed by: M. Carty, 10 April 68
Copy to: SECNAV Files #2852 SECDEF WH# 6638
Copy of l and r tp : CMC Code DNA for "B" Attn: Maj Hogsett
    Request copy of your action

Attachment D

NAVSO 5216/5 (REV. 11-66)
S/N-0104-904-1761

UNITED STATES GOVERNMENT

Memorandum

DEPARTMENT OF THE NAVY
DNA-mls
DATE: **14 MAY 1968**

FROM : Deputy Director of Personnel

TO : Commandant of the Marine Corps

SUBJECT : Circumstances surrounding the recovery of the remains of the late Private First Class Michael J. KELLY, 2342397, U. S. Marine Corps

1. The late Private First Class Kelly died on 15 February 1968. He sustained multiple gunshot wounds to the body and massive tissue damage to the entire body from hostile grenades while on a patrol. Only portions of his remains were recoverable.

2. The recovered portions of the remains were positively identified and consigned to the family for burial, with military honors being provided by the Inspector-Instructor Staff in Lima, Ohio.

3. During an operation being conducted by the 2d Battalion, 7th Marines on 25 March 1968, human remains were discovered unrecognizable and decomposed believed to be those of a Marine. This belief was based on the fact that Marine units had previously operated in this area.

4. These facts were reported to the Company Commander and the Tactical Commander in the field. The Tactical Commander, having prior knowledge of the fact that a Marine was previously killed in the same vicinity and that partial remains were recovered and sent to the next of kin, concluded that these newly discovered remains were those of the same Marine. He concluded that to send these remains to the family who had already buried the partial remains, was inadvisable. He also reasoned that if the remains were buried at a known location they could be recovered later, if necessary. Based on these conclusions, the Tactical Commander ordered the remains buried.

5. The remains were buried on 25 March 1968 fifty meters from the Company Command Post.

6. These facts were reported to superior authorities within the 1st Marine Division who countermanded the previous decision and ordered the remains to be exhumed on 27 March 1968 for proper identification and disposition.

7. The Kelly family was made aware that the initial remains were partial.

8. The newly recovered remains were positively identified and the Kelly family was personally and tactfully made aware of the recovery of additional remains of their son.

Buy U.S. Savings Bonds Regularly on the Payroll Savings Plan

DNA-mls

9. During these trying times, the Kelly family has been completely understanding.

10. In view of the unique and sensitive circumstances of this particular case, TAB A is a proposed letter to be sent to the Kelly family. It is recommended that the Commandant sign the letter at TAB A.

Very respectfully,

WILLIAM K. JONES

DNA-mls
30 Apr 1968

Mr. and Mrs. John H. Kelly

Dear Mr. and Mrs. Kelly:

This is to confirm the information provided you during the visit of Major Shariff, Inspector-Instructor Staff, 6th Ordnance Field Maintenance Corps, Lima, Ohio, on 29 April 1968.

The remainder of your late son's remains were recovered by a patrol inserted into the heavily infested area on 27 March 1968. These remains were immediately evacuated to the Mortuary located in Danang where identification procedures were initiated. These identification procedures included comparison of the physical and dental characteristics of the remains to those in your son's health and dental records, all of which were in agreement.

It is regretted that a delay was encountered in establishing positive identification; however, I am sure you will understand this delay was caused to ensure that the remains recovered were those of your late son.

The remains of your son are now located at the Mortuary, Dover Air Force Base, Dover, Delaware, awaiting your disposition instructions. When these instructions are received the remains will be prepared and transferred at government expense commensurate with your desires.

I sincerely regret the additional anxieties you are experiencing. I hope, however, you will understand that earlier complete recovery of your son's remains was not possible due to the tactical situation and the inaccessibility of the location.

Please accept my deepest sympathy. I hope that the passage of time will somewhat diminish the sorrow you feel from your great loss.

Sincerely,

J. F. HOGSETT, JR.
Major, U. S. Marine Corps
Head, Casualty Section
By direction of the Commandant of the Marine Corps

Blind copy to:
I-I, Lima, Ohio

RE: PFC Michael J. KELLY, 2342397, USMC, deceased

CASE FILE

Attachment F

2 May 1968

Honorable Robert F. Kennedy
United States Senate
Washington, D. C.

Dear Senator Kennedy:

This is in further reply to your inquiry of 2 April 1968 regarding Corporal Willie Williams, 2287883, and your recent letter regarding Private First Class William P. Kelly, 2243492, U. S. Marine Corps.

Both of the above mentioned letters concern the same incident so I have taken the liberty of responding to them both in a single reply.

During a recent operation in Vietnam, members of the 2d Platoon of Company "E", 2d Battalion, 7th Marines, 1st Marine Division, at approximately 10:25 a.m., on 25 March 1968, discovered unrecognizable and decomposed remains believed to be those of a Marine. This belief was based on the fact that Marine units had previously operated in this area.

These facts were reported to the Company Commander and the Tactical Commander in the field. The Tactical Commander, having prior knowledge of the fact that a Marine was previously killed in the same vicinity and that partial remains were recovered and sent to the United States for burial by the next of kin, concluded that these newly discovered remains were those of the same Marine. He concluded that to send these remains to the family, who had already buried the partial remains previously recovered, was inadvisable. He also reasoned that if the remains were buried at a known location they could be recovered later, if necessary. Based on these conclusions, the Tactical Commander ordered the remains buried. The remains were buried in a marked grave by some officers and one noncommissioned officer approximately fifty meters from the Company Command Post, on 25 March 1968. These facts were reported to superior authorities within the 1st Marine Division who countermanded the previous decision and ordered the remains to be exhumed on 27 March 1968 for proper identification and disposition.

The decision to bury the remains on the part of the Tactical Commander reflected his sincere desire to spare the next of kin additional grief and anxiety.

The members of the 2d Platoon of Company "E" were made aware of the exhumation of the remains for proper identification, and burial by the next of kin.

It should be pointed out that the next of kin of the Marine concerned were, in fact, notified that the initial remains were partial. The additional remains have now been interred by the next of kin with the initial remains.

The release of information of this nature is normally subject to rigid control. It is furnished to you as a member of Congress knowing that it will be subject to discretion as to its use and dissemination.

You can be assured that all concerned are well aware of the importance of returning our dead to the United States and to their families for burial with the dignity, honor, and respect they so preeminently deserve.

I trust that the above information is satisfactory for your purposes.

Sincerely,

Lt. General Dulacki (signature that appears on the original letter)

2 Aug 1968

Honorable Robert F. Kennedy
United States Senate
Washington, D. C. 20510

Dear Senator Kennedy:

This is in further reply to your inquiry of 2 April 1968 regarding Corporal Willie WILLIAMS, 2287883, and your recent letter regarding Private First Class William D. KELLY, 2243692, U. S. Marine Corps.

Both of the above mentioned letters concern the same incident so I have taken the liberty of responding to them both in a single reply.

During a recent operation in Vietnam, members of the 2d Platoon of Company "F", 2d Battalion, 7th Marines, 1st Marine Division, at approximately 10:25 a.m., on 25 March 1968, discovered unrecognizable and decomposed remains believed to be those of a Marine. This belief was based on the fact that Marine units had previously operated in this area.

These facts were reported to the Company Commander and the Tactical Commander in the field. The Tactical Commander, having prior knowledge of the fact that a Marine was previously killed in the same vicinity and that partial remains were recovered and sent to the United States for burial by the next of kin, concluded that these newly discovered remains were those of the same Marine. He concluded that to send these remains to the family, who had already buried the partial remains previously recovered, was inadvisable. He also reasoned that if the remains were buried at a known location they could be recovered later, if necessary. Based on these conclusions, the Tactical Commander ordered the remains buried. The remains were buried in a marked grave by some officers and one noncommissioned officer approximately fifty meters from the Company Command Post, on 25 March 1968.

NOTE: THE REMAINS RECOVERED AS MENTIONED IN THIS LTR WERE THOSE OF Michael J. Kelly. THE Kelly MENTIONED IN PARA 1 IS NO RELATION TO THE Deceased. THIS IS FILED IN THIS CASE FOR CROSS REFERENCE PURPOSES ONLY - DO NOT REMOVE. DNA

These facts were reported to superior authorities within the 1st Marine Division who countermanded the previous decision and ordered the remains to be exhumed on 27 March 1969 for proper identification and disposition.

The decision to bury the remains on the part of the Tactical Commander reflected his sincere desire to spare the next of kin additional grief and anxiety.

The members of the 2d Platoon of Company "E" were made aware of the exhumation of the remains for proper identification and burial by the next of kin.

It should be pointed out that the next of kin of the Marine concerned were, in fact, notified that the initial remains were partial. The additional remains have now been interred by the next of kin with the initial remains.

The release of information of this nature is normally subject to rigid controls. It is furnished to you as a member of Congress knowing that it will be subject to discretion as to its use and dissemination.

You can be assured that all concerned are well aware of the importance of returning our dead to the United States and to their families for burial with the dignity, honor, and respect they so preeminently deserve.

I trust that the above information is satisfactory for your purposes.

Sincerely,

Attachment H
Author is standing to the far right.

PLEASE CREDIT
U.S. MARINE CORPS PHOTO RELEASE
BY
M.M. UPTON
1ST MARDIV
MAR 30 1968   FIELD NO.   2 90-68

Gen. Robertson — 1st Mar. Div. C.O.

Joe Alsop — Syndicated Columnist - Washington, D.C.

Col. Mueller — 2nd Bn., 7th Mar. C.O.

1/Lt. Stair — E Co., 2nd Bn., 7th Mar. Exec. Off.

Arriving for briefing at E Co.

Combat Base on Hill 190.

Attachment I

Attachment J

Moving to top of Hill 190 for briefing of tactical situation in E-2/9 T.A.O.R. and E Co. appraisal of Operation Worth.

Attachment L
Back Row: John Brown, III, Dr. Donald Catherall, Doug Chamberlain, Col. Jon M. Lauder
Front Row: Sandra Berger, Doug Berger

Attachment M
Sandra Berger, Doug Berger, Shawn Meagley

Attachment O
Sandra Berger, Doug Berger, Doug Chamberlain, Shawn Meagley

Attachment P
Col. Jon M. Lauder, Doug Chamberlain, John Brown, III

Acknowledgements

COOL WATER. Words and Music by BOB NOLAN. Copyright © 1936 (Renewed) MUSIC OF THE WEST (c/o THE SONGWRITERS GUILD OF AMERICA) in the U.S. Canadian Rights Controlled by UNICHAPPELL MUSIC, INC. Exclusive Print Rights Administered by ALFRED MUSIC PUBLISHING. All Rights Reserved. Used By Permission of ALFRED MUSIC

Semper Fi. Words and Music by Kenny Beard, Trace Adkins and Monty Criswell © 2011 Beardtown Music/Tunes of Swan Song Publishing (ASCAP) Exclusive Worldwide Print Rights for Beardtown Music/Tunes of Swan Song administered by Do Write Music, LLC All Rights Reserved. Used by Permission.

Semper Fi. Words and Music by Monty Criswell, Trace Adkins and Kenny Beard. Copyright (c) 2011 Sony/ATV Music Publishing LLC, Spirit Catalog Hold ings S.A.R.L. and Co-Publishers. All Rights on behalf of Sony/ATV Music Publishing LLC Administered by Sony/ATV Music Publishing LLC, 424 Church Street, Suite 1 200, Nashville TN 37219. All Rights on behalf of Spirit Catalog Holdings S.A.R.L. Administered by Spirit Two Nashville and Spirit Of Nashville One International Copyright Secured. All Rights Reserved. Reprinted by Permission of Hal Leonard LLC.

Semper Fi. Words by Kenny Beard, Trace Adkins and Monty Criswell. © 2011 Trace Adkins (ASCAP). Exclusive Worldwide Print Rights for Trace Adkins administered by Fox Rothchild LLP. All Rights Reserved. Used by permission.

Without the steadfast support and encouragement of Grady T. Birdsong, who served as a Marine Corporal in the the First Battalion, 27th Marines during part of the time period covered in this book, my Memoirs would not have ever been completed. After I made his acquaintance in early 2019, he embraced me as a Marine Brother, directed my efforts in self-publishing, and was the culminating effort that led me to complete my literary effort. I will always cherish his friendship and I cannot adequately describe my admiration of him.

Virginia Cusack is one of the most talented copy editors that I have ever known. The regime she maintained as she scoured the details of my manuscript went beyond all of my expectations. I could not have achieved the finality of my memoirs without her.

Nick Zelinger of NZGraphics provided countless hours of work to design the maps included herein. He is a man of immense character and understanding, and I appreciate his dedication to the tasks that brought visual clarity to the many references in this book.

And finally, words cannot adequately express my appreciation of Jan and Joe McDaniel, owners of Bookcrafters in Parker, Colorado; Jan for her tireless efforts to prepare my manuscript for publication, and Joe for his professional design of this book cover. Together, this dedicated team provided the publishing services for me and catapulted my dreams into a reality. Time will never erase my appreciation for and the admiration of these marvelous people.

www.ingramcontent.com/pod-product-compliance
Lightning Source LLC
Chambersburg PA
CBHW030050100526
44591CB00008B/80